Advanced Topics in Science and Technology in China

Volume 61

Zhejiang University is one of the leading universities in China. In Advanced Topics in Science and Technology in China, Zhejiang University Press and Springer jointly publish monographs by Chinese scholars and professors, as well as invited authors and editors from abroad who are outstanding experts and scholars in their fields. This series will be of interest to researchers, lecturers, and graduate students alike.

Advanced Topics in Science and Technology in China aims to present the latest and most cutting-edge theories, techniques, and methodologies in various research areas in China. It covers all disciplines in the fields of natural science and technology, including but not limited to, computer science, materials science, the life sciences, engineering, environmental sciences, mathematics, and physics.

This book series is indexed by the SCOPUS database.

If you are interested in publishing your book in the series, please contact Dr. Mengchu Huang(Email: mengchu.huang@springer.com).

More information about this series at http://www.springer.com/series/7887

Jianhua Yao · Bo Li · Liang Wang

Advanced Laser Process for Surface Enhancement

ZHEJIANG UNIVERSITY PRESS
浙江大学出版社

Springer

Jianhua Yao
College of Mechanical Engineering
Zhejiang University of Technology
Hangzhou, China

Bo Li
Institute of Laser Advanced Manufacturing
Zhejiang University of Technology
Hangzhou, China

Liang Wang
Institute of Laser Advanced Manufacturing
Zhejiang University of Technology
Hangzhou, China

ISSN 1995-6819 ISSN 1995-6827 (electronic)
Advanced Topics in Science and Technology in China
ISBN 978-981-15-9658-2 ISBN 978-981-15-9659-9 (eBook)
https://doi.org/10.1007/978-981-15-9659-9

Jointly published with Zhejiang University Press, China
The print edition is not for sale in China Mainland. Customers from China Mainland please order the print book from: Zhejiang University Press.

This Springer imprint is published by the registered company Springer Nature Singapore Pte Ltd.
The registered company address is: 152 Beach Road, #21-01/04 Gateway East, Singapore 189721, Singapore

Preface

In recent years, with the rapid development of high-end manufacturing, the technologies of laser surface modification and improvement have received extensive application for performance enhancement of various key parts or components in the fields of energy, marine, chemical and aerospace engineering, achieving huge economic and social profit. In the meanwhile, with continuous improvement and innovation in manufacturing technologies, the corresponding laser surface modification techniques consequently exhibit growing trend in both speed and diversity, changing from single process to multiple or compound processes which not only retain the features of the former, but also take the advantages of the latter, with the assistance of other energy fields to gain better processing effectiveness that single process cannot attain. This has significantly expanded the application of laser surface modification technologies in various industries.

Two typical hybrid laser surface modification processes, so-called electro/magnetic field aided laser process and supersonic laser deposition process, are introduced in the book, which enable solving the common problems in quality control and low efficiency of single laser surface modification process, such as severe contamination and high consumption. The contents of this book include recent research outcomes of the authors' group, in various aspects such as numerical modeling, theoretical analysis, experimental data, etc., presented with the assistance of figures and tables, so as to help readers understand the two advanced laser processes easier.

From this book, the readers can learn more knowledge about the principle, technique, and application of electro/magnetic field aided laser and supersonic laser deposition technologies, and understand the means of controlling the laser process to achieve high performance by combining laser with ultrasonic and electric/magnetic field, thus to break through the bottlenecks of existing single laser processing techniques. The book is aimed to provide the guidance and reference for the researchers, engineers, and students in the fields of mechanical engineering, laser processing, and material engineering.

This book is written by Prof. Jianhua Yao, Associated Professor Bo Li and Liang Wang, Zhejiang University of Technology, China. Associated Professor Qunli Zhang, Dr. Zhijun Chen, Dr. Honghao Ge, Dr. Lijing Yang, Dr. Zhihong Li, Dr. Yong

Hu, and Dr. Lijuan Wu have also contributed to the research and relevant experiments. Professor Rong Liu from Carleton University, Canada, has edited this book in English. For systematical integrity, this book has cited the research results published in book, journal, and conference articles from international and domestic experts in the related fields, hence, we wish to express sincere thanks to them. The research projects covered in this book have been sponsored and supported by the National Key R&D Program of China (2017YFB1103600, 2018YFB0407300), and the National Natural Science Foundation of China (U1509201, 51701182, 51705464, 51804274). Here, we all express our great appreciation.

There are inevitably mistakes and errors in the book due to our oversight, please feel free to make comments and criticization.

Hangzhou, China

Jianhua Yao
Bo Li
Liang Wang

Contents

Chapter 1
Introduction

Abstract In this chapter, the fundamental concepts of laser technologies and applications of laser in surface engineering are briefly introduced. Various laser processes with associated parameters are reviewed. The influences of individual process parameters chosen on the final laser processed products are generally discussed. The background and significance of the newly developed laser processes which are detailed in this book are delineated. These advanced processes include incorporating a steady magnetic field in laser remelting process and an electric magnetic compound field in laser melt injection process, a novel deposition technique, known as supersonic laser deposition, by introducing laser into cold spray process, that is, the deposition site of cold spray is simultaneously heated by laser in order to preheat and soften the spraying particles and substrate.

1.1 Laser Application in Surface Engineering

1.1.1 Laser Processing of Materials

Invented in 1960, laser offers a unique set of opportunities for precise delivery of high-quality coherent energy. This unique behavior results in very good focusing capabilities, and the resulting power densities in the focus of a high-power laser can reach 1000 W/cm^2 easily [1]. Such high intensities enable rapid thermal processing, where thermal energy is confined to the beam spot without collateral damage to the adjacent material [2]. Nowadays laser systems can be found in nearly all branches of industry for numerous applications. This is a result among other things of close interactions between the industrial needs and fundamental research in systems, materials and processes performed on a scientific and technical basis. Laser processing covers a wide range of power, interaction time and process materials with length scales from nanometers to meters [3].

J. Yao et al., *Advanced Laser Process for Surface Enhancement*, Advanced Topics in Science and Technology in China 61, https://doi.org/10.1007/978-981-15-9659-9_1

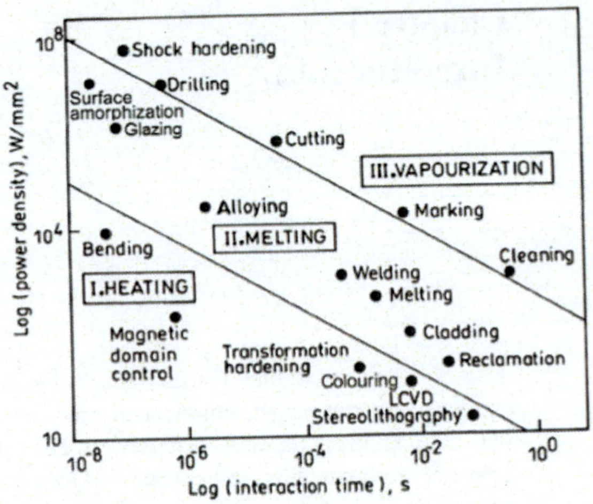

Fig. 1.1 Classification of different laser processing techniques [5]

Material processing is one of the important and active areas of research in heat transfer today. Rapid thermal processing of materials refers to manufacturing and material fabrication techniques that are strongly dependent on the thermal transport mechanisms, involving rapid heating and cooling processes [4]. Laser material processing can be divided into three major classes, namely heating, melting and vaporisation. These three classes are determined by power density (irradiance) and exposure/interaction time (residence time), as illustrated by the chart in Fig. 1.1 [5]. Within these classes there are various processes that can be achieved through careful selection of the irradiance and residence time.

Laser surface treatment of materials offers considerable advantages over the conventional methods. It is an important technique because it offers a possibility to enhance various properties such as the surface strength, hardness, roughness, coefficient of friction, chemical resistance, and corrosion of various materials. Such improvements to a material surface are not only ideal for applications when wear rate and shear stresses are high but could also be used for maintaining or elongating the component functional life by means of reducing the microcracks in surfaces. In addition, aesthetics can also be improved using laser surface treatment (for ceramics in particular) by creating a modified surface layer.

1.1.2 *Various Laser Processes*

For the laser applications in surface engineering, there are various surface treatment techniques, among which laser glazing, laser melting, laser alloying, laser cladding, and laser shock hardening, are the most popular.

The term 'glazing' refers to the process of being able to make 'glassy', to make a solid without any crystalline structure [6]. Laser glazing occurs when a beam of sufficient intensity, to create a molten state, is scanned rapidly across a solid. Extremely rapid cooling caused by conduction of heat into the bulk material and shallow processed area results in major microstructural alteration, that is, freezing of the atoms in a random state before they can form crystals. These microstructural changes enable laser glazing to produce high compressive strength, hardness, wear and corrosion resistance properties [7]. The reduction of grain sizes into nanoscale can increase strength, hardness and tribological properties of both metals and ceramics [8]. An amorphous structure has no grains and is known to increase hardness and surface properties through elimination of crystalline anisotropy and inter-crystalline defects [6].

In laser melting process the alloy or steel surface is laser melted and rapidly re-solidified without any direct addition of other material elements to modify the chemical composition of the surface [9]. Vaporization is avoided during this laser treatment and the molten pool solidifies rapidly. There is a sequence of events which initiate when the laser beam contacts the target surface. Firstly, the near surface region rapidly reaches the melting point and then a liquid/solid interface starts to move through the alloy. Diffusion of elements follows within this liquid phase. At this point the laser pulse is nearly terminated while the surface has remained below the vaporization temperature. The maximum melt depth has been attained, inter-diffusion continues, and the re-solidified interface velocity is momentarily zero at this stage. The interface then rapidly moves back to the surface from the region of maximum melt depth. Inter-diffusion continues in the liquid, but the re-solidified metal behind the liquid/solid interface cools so rapidly that solid state diffusion is negligible compared to that expected from an equilibrium phase diagram. Finally, re-solidification is completed and a surface modified layer has been created. Due to the relatively small depth of the melted zone (10–1000 μm), very high quench rates are achieved, resulting in nonequilibrium martensitic microstructure [10].

Laser alloying is a surface modification technique which utilizes high power density available from focused laser sources to melt, externally added, alloying elements and the underlying substrate [11]. This process is similar to laser melting except that extraneous alloying elements are added to the melt pool, thus changing the chemical composition of the surface. There are two main ways used to deposit alloying elements, co-deposition and pre-deposition. Co-deposition involves direct deposition of alloying elements during laser irradiation. The alloying elements are deposited mainly in the form of gas [12], powder [13] or wire [14]. Co-deposition of alloying elements is considered the most attractive as it involves a single step processing. This method also offers the significant advantage of real time control over pre-deposition where the alloying elements are deposited prior to laser irradiation.

Laser cladding is a melting process where there is fusion of a substrate material with a material of different metallurgical properties. Laser cladding is similar to laser surface alloying except the dilution by the substrate is kept at a minimum. Normally, the processes involving less than 10% dilution are referred to as cladding while those with dilutions exceeding 10% are termed as alloying [15]. The cladding process offers optimum bonding, great flexibility, low distortion, and low thermal load on the workpiece together with little need for post cladding treatment [16].

However, the main drawback of cladding is that it often produces uncontrollable cracks, which result from the very high cooling rate of the melted pool.

Laser shock hardening is to use a high intensity laser and suitable overlays to generate high pressure shock waves on the surface of the treated material. When a material is irradiated with a Q switched pulse laser, operated at power density greater than 0.1 GW/cm^2, a shock wave can be generated on the surface [17]. Surface of the treated material can be coated with a black paint before laser shock hardening as a laser energy absorbing layer. The workpiece is usually covered with a transparent layer (water or glass). When the laser beam strikes the coated sample, the black paint, due to absorption of laser beam energy, is heated and instantaneously vaporized. The vapour absorbs the remaining laser beam radiation and then produces plasma. The rapidly expanding plasma creates a high surface pressure, which propagates into material as a shock wave. The shock wave can induce compressive residual stress that penetrates beneath the surface and strengthens the surface of the treated material. This effect may induce microstructural changes, cause a high increase in dislocation density, and influence the surface roughness of the material as well as introduce a compressive residual stresses into the treated surface of the material [18]. Although laser shock hardening improves the wear and corrosion properties, the improvements are insignificant compared to other laser surface modification techniques.

1.2 Laser Process Parameters

1.2.1 Influence Factors of Laser Process

The main process variables in the laser surface modification operation include laser power density, laser beam diameter and configuration, laser scan rate, etc.[19]. In order to obtain the required properties of the surface it is crucial that the selected interaction time and power density allow the material to undergo the desired degree of heating and phase transition. Power density (irradiance) input determines the maximum temperature attained. Duration of interaction (residence time) governs the reaction among the phases and the cooling rates. Interfacial properties of the treated surface layer are also strongly attributed to the temperature encountered during treatment [20]. The control of these laser processing parameters is important to achieve repeatability and optimum microstructure changes on the surface. It should be noted that the parameter selection process is complex. For instance, it would be more convenient to quantify different laser processes using a singular parameter like energy density determined from power density multiplied by time, but this would not define the true outcome of each individual processing parameter as it is the specific combination of power and time (rather than their product) which determines the resulting thermal and material effect.

1.2.2 Laser Operating Modes

Both continuous wave and pulsed mode laser may be employed for surface modification. In contrast to the continuous wave operation, a pulsed beam offers several challenges owing to the higher number of operating variables and the complexity of optimising the process parameters. For a given melt profile, surface characteristics and other aspects including economy, the operating parameters such as pulse energy, pulse width, pulse frequency, scan rate etc. have to be chosen. This calls for an excellent knowledge of the influence of operating parameters on melt profile and an in depth understanding of its effect on structure and properties of the treated surface [21]. Compared with continuous wave lasers, the high peak power of a single pulse produced by a pulsed laser can readily induce higher absorptivity in metals. Short duration time of pulsed laser means short heating time which therefore contributes to higher cooling rate, subsequent grain refinement, and the formation of non-equilibrium phases in the treated surface [22]. Since pulsed laser allows for adjustment of various pulse parameters, it can provide a flexible processing window which can be tailored to suit certain microstructures and materials [23]. Continuous wave operation on the other hand can offer higher depth of processing compared to pulsed mode [24].

1.2.3 Laser Power and Irradiance

The average power of the laser beam, whether it operates in the continuous mode or pulsed mode, affects the nature and scale of the resultant surface [25]. The use of pulse power allows two more variables to be concerned, namely, pulse repetition frequency and percentage overlap. Penetration of the laser beam is a function of power. However, high power setting results in vaporisation and material ejection [26]. The power density or irradiance of a laser beam is determined by the power of the laser beam used and the diameter of its focal spot (spot size). Figure 1.2 illustrates the laser beam profile depending on the focusing.

Many studies have focused on the beam either below or above the surface to achieve larger beam diameters than the focused spot size and hence give the power density required for a specific process [27–29]. Generally, the laser intensity along with the material properties of the surface gives an indication of the expected rise in temperature in the focal region and consequently indicates the melting intensity of the material [30].

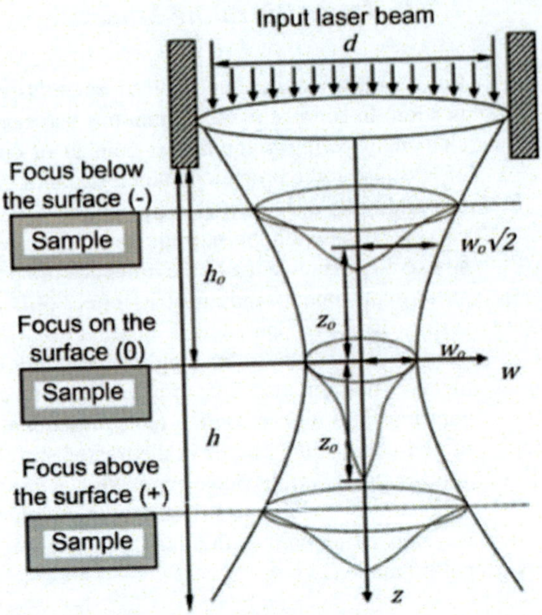

Fig. 1.2 Geometry and intensity of the Gaussian laser beam [27]

1.2.4 Laser Scanning Rate and Residence Time

The rate of laser scanning and laser beam spot size are crucial in determining the residence time of the heat at the target surface. Residence time can be defined as the amount of time when the laser beam is in contact with the coating material. For pulsed mode, residence time encompasses the duty cycle, laser scanning rate and beam diameter. This time is equivalent of exposure time for continuous mode laser [31]. In general, the width and height of the cladding layer decrease with an increase in laser scanning speed under comparable laser cladding conditions because the amount of powder fed into the molten pool per unit length is reduced as the scanning speed increases. On the other hand, increasing scanning speed leads to an increase in the depth of the melted substrate, and in turn, the dilution rate. This can be attributed to the weaker "heat shielding effect" [32] of the powders and more heat absorbed by the substrates at higher scanning speeds.

Energy density is the power per unit time per unit area delivered by a single laser pulse at the focal spot. It is a more accurate parameter to compare the effect of a laser beam with varying pulse durations. Energy density is the product of irradiance and residence time.

For multiple tracks of laser cladding, overlap is the ratio of the distance between two consecutive laser spots and the diameter of the spot. A homogeneous structure can be achieved in pulsed laser beams by using controlled overlap.

1.3 Advanced Laser Process

1.3.1 Laser Process Optimization

To achieve desired surfaces that are treated with laser, it is necessary to use optimal process parameters. In practice, obtaining optimal process parameters is normally made by repeating the laser process with varying the process parameters, toward the target performance of the surface, microstructure, mechanical, tribological and corrosion properties. For instance, as one of the surface modification techniques, laser remelting allows selective heating and melting of a surface, and modifies the surface properties by rapid melting and solidification [33]. Due to high cooling rate of the surface layer, the resulting microstructure of the surface layer is very different from the substrate metal, leading to improved properties in various aspects such as better wear, corrosion, and thermal fatigue resistance [34]. However, surface undulation after laser remelting is always a problem because it degrades the surface quality [35]. To solve this problem, the common way is to optimize the process parameters, for example, by using different laser power densities, laser scanning rate, laser spot size, etc., then examining the surface morphology with scanning electron microscope (SEM). In general, this approach can somewhat improve the surface undulation, but since the process parameters not only influence the surface morphology, but also affect other properties of the surface such as wear and corrosion resistance, improvement on the surface morphology may compromise with detriment of other important properties of the surface. Additionally, the window for adjusting the process parameters is usually limited by the equipment capacities, therefore, relying on optimization of process parameters to solve the surface undulation is not an ideal means.

Another laser application in surface modification is creating metal matrix composites (MMCs) reinforced with ceramic particles via laser melt injection process. The MMC coating layer prepared from this process on a metallic substrate exhibits special characteristics such low particle dissolution rate, high surface performance and low cracking tendency [36]. Unlike laser cladding process, during laser melt injection process the reinforcement particles (usually ceramics) are injected in the molten pool without any other metal–matrix powders and move with the melt flow, thus preserve solid state or micro melt state due to the rapid solidification [37]. For the coating layers prepared with laser melt injection, the graded distribution of reinforcement particles is the main concern, which can be designed at a microstructural level to tailor specific materials for their functional performance in particular applications [38]. To achieve this goal, traditional means is to control the key process parameters which are related to the powder injection such as the injection angle with respect to the surface normal, the relative position between the powder spot and laser spot and powder injection velocity [39], because these parameters influence the distribution of reinforcement particles significantly. However, it has been noted that practical implementation of the traditional approach has difficulties, for example, it is time-consuming and also the adjustable window for lateral nozzle powder delivery system is very narrow [40].

1.3.2 Laser Process with Auxiliary Device

Instead of improving the surface undulation in laser remelting and controlling the reinforcement particle distribution in laser melt injection by optimizing the process parameters, an advanced approach to achieving these goals is introduced in this book, which incorporates a steady magnetic field in laser remelting process [41] and an electric–magnetic compound field in laser melt injection process [42, 43].

In the laser remelting process of low carbon low alloy steel ASTM A529M, a 2 kW Laserline diode laser was coupled with a steady magnetic field which was created by four NdFeB permanent magnets. When a steady magnetic field was applied, the surface undulation of the remolten steel could be suppressed by dissipating effect on the molten pool dynamics [41]. From experimental and modelling studies in the influences of temperature and velocity distributions on the molten pool under a magnetic field applied, it was found that the main influence of the steady magnetic field on the molten pool was the damping effect of the magnetic field induced by the Lorenz force in the opposite direction of melt flow, which suppressed significantly the overall velocity of the molten metal flow and the internal double vortices. The surface undulation of the remolten specimen was then reduced greatly due to the decrease of the normal melt flow velocity.

An electric–magnetic compound field was utilized in the laser melt injection process of WC particles on AISI 316L austenitic stainless steel, which was coupled with a 2 kW Laserline diode laser and applied to to the molten pool [42, 43]. The steady magnetic field was provided by using electromagnets and the steady electric field was created by employing large capacity lead-acid batteries. The experimental and numerical investigations revealed that with an electric–magnetic compound field applied, both the induced Lorentz force and directional Lorentz force were generated in the molten pool. The induced Lorentz force provided the suppression effect on the velocity of the molten pool and the directional Lorentz force could change the equivalent buoyancy acting on the particles, which was the main factor controlling the distribution of the reinforcement particles.

1.4 Laser Contribution to Cold Spray

1.4.1 Cold Spray of Hard Materials

Cold spray, as a relatively new material deposition technique, has attracted much attention in recent years, owing to many unique features such as low risk of high temperature oxidation, evaporation, melting, residual stresses and other common problems inherent to traditional thermal spraying [44]. In a cold spray process, metal powders are accelerated by a supersonic gas flow to a high velocity ranging from 300 to 1200 m/s and then impinge onto the substrate or already deposited coating at a temperature well below the melting point of deposited materials [45]. The high-velocity impact can induce intensive plastic deformation of the spraying particles and the substrate, enabling the formation of a coating.

This deposition means has been reported to work well for soft materials such as copper and aluminum. However, it is not suitable for hard materials. Because the formation of coating via cold spray relies solely on the kinetic energy rather than the combined effects of both kinetic and thermal energies available in conventional thermal spraying, hard metals generally exhibit little plastic deformation [46]. Therefore, any attempt in enhancing the plastic deformation of deposited material and substrate will favor the bonding of the spraying particles to the substrate. As a common sense, materials can be softened by heating to a high temperature and they display larger plastic deformation when they become softer.

1.4.2 Laser Aided Cold Spray Process

In recent years, a novel deposition technique, known as supersonic laser deposition, has been developed by introducing laser into cold spray process [47]. In this process, the deposition site of cold spray is simultaneously heated by laser in order to preheat and soften the spraying particles and substrate. Heating the particles not only avoids the nozzle fouling but also significantly reduces the critical velocity for deposition of the particles, which allows bonding to occur upon impact at the velocity of spraying particles about half that in cold spray process. Because of these unique features, supersonic laser deposition has been applied on various materials, in particular, hard materials such as Stellite alloys, Ni60, diamond, and so on [48–50].

In this book, supersonic laser deposition technique is introduced and applied on a wide range of materials from soft metal copper to hard Stellite 6 and nickel alloy Ni60 as well as very hard diamond and composites [48–53]. For the soft metal copper, when deposited on a medium-carbon steel via supersonic laser deposition process, it was found that the coating surface was much improved in morphology with laser irradiation and the deposition efficiency was also increased significantly. The deposited coating was denser and showed the good coating/substrate interfacial bonding [48]. For the hard alloys, Stellite 6 and Ni60, the experimental investigations revealed significant improvement of laser irradiation addition on the surface undulation and deposition efficiency of the Stellite 6 and Ni60 deposited on a medium-carbon steel in the cold spray process. The Stellite 6 and Ni60 coating prepared via supersonic laser deposition retained the original microstructure and phases of the feedstock material due to relatively low laser energy input. Compared with the Stellite 6 and Ni60 coating prepared via laser cladding, they exhibited the superior wear/corrosion-resistant properties [49, 50].

With application of supersonic laser deposition technique, diamond/Ni60, WC/SS316L and WC/Stellite 6 composite were deposited successfully on a medium carbon steel or stainless steel, which was unable to do with cold spray solely [51–53]. It was found that the diamond particles were uniformly distributed in the Ni60 matrix and no visible cracks were present in the coating. During the supersonic laser deposition process graphitization transition was suppressed. The coating exhibited dense microstructure and excellent wear resistance [51]. Compared with the cold spray, the WC/SS316L composite coating produced from the supersonic laser deposition process had higher deposition efficiency, WC concentration and interface bonding, which improved the tribological properties of the coating [52]. The comparison between the WC/Stellite 6 composite coatings prepared via supersonic laser deposition and laser cladding revealed that the former had less defects, higher WC content, lower cracking susceptibility and better wear resistance. These good properties were all attributed to the retention of microstructure and phase compositions of the feedstock materials in the supersonic laser deposition process because of relatively low heat involved. In the meanwhile, additional heat from laser changed the interface bonding mechanism of the coating from dominant mechanical bonding in the cold spray coating to coexistence of mechanical and metallurgical bonding, which enhanced the coating interface bonding [53].

References

1. I. Smurov, M. Doubenskaia, S.N. Grigoriev, D.V. Kotoban, P.A. Podrabinnik, Application of laser surface engineering to solve tribological problems. J. Frict. Wear **35**(6), 470–476 (2014)
2. E. Chikarakara, S. Naher, D. Brabazon, Process mapping of laser surface modifucation of AISI 316L stainless steel for biomedical applications. Appl. Phys. A Mater. Sci. Process. **101**(2), 367–371 (2010)
3. M.D. Wang, S.H. Shi, D.W. Zuo, M. Wang, Study on variable spot cladding for large area laser surface modifucation. Adv. Mater. Res. **53–54**, 319–324 (2008)
4. G.Z. Huang, Z.H. Li, J.Y. Tang, Effect of laser surface modification and tempering process on microstructure and hardness profiles of roll materials. Adv. Mater. Res. **314–316**(3), 1900–1905 (2011)
5. J.D. Majumdar, I. Manna, Laser processing of materials. Sadhana **28**(3–4), 495–562 (2003)
6. H.L. Tsai, P.C. Tsai, Laser glazing of plasma-sprayed zirconia coatings. J. Mater. Eng. Perform. **7**(2), 258–264 (1998)
7. D.T.A. Matthews, V. Ocelík, J.T.M. de Hosson, Tribological and mechanical properties of high power laser surface-treated metallic glasses. Mater. Sci. Eng. A **471**(1–2), 155–164 (2007)
8. A.A. Peligrad, E. Zhou, D. Morton, L. Li, A melt depth prediction model for quality control of laser surface glazing of inhomogeneous materials. Opt. Laser Technol. **33**(1), 7–13 (2001)

9. M.R. Alkahari, T. Furumoto, T. Ueda, A. Hosokawa, R. Tanaka, M.S. Abdul Aziz, Thermal conductivity of metal powder and consolidated material fabricated via selective laser melting. Key Eng. Mater. 523–524 (2012) 244–249
10. D.B. Witkin, D.N. Patel, G.E. Bean, Notched fatigue testing of Inconel 718 prepared by selective laser melting. Fatigue Fract. Eng. Mater. Struct. **42**(1), 166–177 (2019)
11. C.W. Draper, C.A. Ewing, Laser surface alloying: a bibliography. J. Mater. Sci. **19**(12), 3815–3825 (1984)
12. J. Liu, Q. Luo, Z. Zou, Laser gas alloying of titanium alloy with nitrogen. Surf. Coat. Technol. **57**(2–3), 191–195 (1993)
13. E. Brandl, F. Palm, V. Michailov, B. Viehweger, C. Leyens, Mechanical properties of additive manufactured titanium (Ti-6Al-4V) blocks deposited by a Solid-State Laser and Wire. Mater. Des. **32**(10), 4665–4675 (2011)
14. H.C. Man, M. Bai, F.T. Cheng, Laser diffusion nitriding of Ti-6Al-4V for improving hardness and wear resistance. Appl. Surf. Sci. **258**(1), 436–441 (2011)
15. R.L. Sun, J.F. Mao, D.Z. Yang, Microstructural characterization of NiCr1BSiC laser clad layer on titanium alloy substrate. Surf. Coat. Technol. **150**(2–3), 199–204 (2002)
16. Y. Wang, H.M. Wang, Wear resistance of laser clad Ti_2Ni_3Si reinforced intermetallic composite coatings on titanium alloy. Appl. Surf. Sci. **229**(1–4), 81–86 (2004)
17. A. Kruusing, Underwater and water-assisted laser processing: Part 2 - Etching, cutting and rarely used methods. Opt. Laser Eng. **41**(2), 329–352 (2004)
18. M. Rozmus-Gornikowska, Surface modifications of a Ti6A14V alloy by a laser shock processing. Acta Phys. Polomica Ser. A **117**(5), 808–811 (2010)
19. G. Demian, M. Taca, I. Gutu, M. Demian, Effect of laser process parameters on the surface absorptivity. Mater. Sci. Forum **480–481**, 207–211 (2005)
20. L. Mugwagwa, D. Dimitrov, S. Matope, I. Yadroitsev, Influence of process parameters on residual stress related distortions in selective laser melting. Procedia Manuf. **21**, 92–99 (2018)
21. V.C. Kumar, Process parameters influencing melt profile and hardness of pulsed laser treated Ti-6Al-4V. Surf. Coat. Technol. **201**(6), 3174–3180 (2006)
22. A.N. Samant, B. Du, S.R. Paital, S. Kumar, N.B. Dahotre, Pulsed laser surface treatment of magnesium alloy: correlation between thermal model and experimental observations. J. Mater. Process. Technol. **209**(11), 5060–5067 (2009)
23. A.J. Pinkerton, L. Li, An investigation of the effect of pulse frequency in laser multiple layer cladding of stainless steel. Appl. Surf. Sci. **208–209**, 405–410 (2003)
24. A.N. Samant, S.P. Harimkar, N.B. Dahotre, Laser beam operation mode dependent grain morphology of alumina. J. Appl. Phys. **102**(12), 1–6 (2007)
25. N. Zaveri, M. Mahapatra, A. Deceuster, Y. Peng, L. Li, A. Zhou, Corrosion resistance of pulsed laser-treated Ti-6Al-4V implant in simulated biofluids. Electrochim. Acta **53**(15), 5022–5032 (2008)
26. Z. Sun, I. Annergren, D. Pan, T.A. Mai, Effect of laser surface remelting on the corrosion behavior of commercially pure titanium sheet. Mater. Sci. Eng. A **345**(1–2), 293–300 (2003)
27. E.V. Bordatchev, S.K. Nikumb, Effect of focus position on informational properties of acoustic emission generated by laser–material interactions. Appl. Surf. Sci. **253**(3), 1122–1129 (2006)
28. E. Ramous, A note on rapid surface melting by laser. J. Mater. Process. Technol. **57**(1–2), 1–3 (1996)
29. C.H. Chen, C.P. Ju, J.M. Rigsbee, Laser surface modification of ductile iron: I. microstructure. Mater. Sci. Technol. 4(2) (1988) 161–166
30. M.S. Trtica, B.M. Gakovic, T.M. Nenadovic, M.M. Mitrovic, Surface modification of stainless steels by TEA CO2 laser. Appl. Surf. Sci. **177**(1–2), 48–57 (2000)
31. J.L. Chen, J. Li, R. Song, L.L. Nai, J.Z. Shao, C.C. Qu, Effect of the scanning speed on microstructural evolution and wear behavior of laser cladding NiCrBSi composite coatings. Opt. Laser Technol. **72**, 86–99 (2015)
32. P. Xu, C.X. Lin, D.P. Sun, The effect of scanning velocity on AISI 304 stainless steel laser cladding coatings. Adv. Mater. Res. **591–593**, 1098–1101 (2012)

33. X. Tong, M.J. Dai, Z.H. Zhang, Thermal fatigue resistance of H13 steel treated by selective laser surface melting and CrNi alloying. Appl. Surf. Sci. **271**, 373–380 (2013)
34. P.H. Chong, Z. Liu, P. Skeldon, P. Crouse, Characterisation and corrosion performance of laser-melted 3CR12 steel. Appl. Surf. Sci. **247**, 362–368 (2005)
35. E. Chikarakara, S. Naher, D. Brabazon, Spinodal decomposition in AISI 316L stainless steel via high-speed laser remelting. Appl. Surf. Sci. **302**, 318–321 (2014)
36. J.A. Vreeling, V. Ocelík, Y.T. Pei, D.T.L. van Agterveld, J.T.M. De Hosson, Laser melt injection in aluminum alloys: on the role of the oxide skin. Acta Mater. 48 (2000) 4225–4233
37. Y.T. Pei, V. Ocelik, J.T.M. De Hosson, SiCp/Ti6Al4V functionally graded materials produced by laser melt injection. Acta Mater. **50**, 2035–2051 (2002)
38. D. Liu, L. Li, F. Li, Y. Chen, WCp/Fe metal matrix composites produced by laser melt injection. Surf. Coat. Technol. **202**, 1771–1777 (2008)
39. J.A. Vreeling, V. Ocelík, J.T.M. De Hosson, Ti–6Al–4V strengthened by laser melt injection of WCp particles. Acta Mater. **50**, 4913–4924 (2002)
40. A.M. Do Nascimento, V. Ocelík, M.C.F. Ierardi, J.T.M. De Hosson, Microstructure of reaction zone in WCp/duplex stainless steels matrix composites processing by laser melt injection. Surf. Coat. Technol. 202 (2008) 2113–2120
41. L. Wang, J.H. Yao, Y. Hu, S.Y. Song, Suppression effect of a steady magnetic field on molten pool during laser remelting. Appl. Surf. Sci. **351**, 794–802 (2015)
42. L. Wang, J.H. Yao, H. Yong, Q.L. Zhang, Z. Sun, R. Liu, Influence of electric-magnetic compound field on the WC particles distribution in laser melt injection. Surf. Coat. Technol. **315**, 32–43 (2017)
43. Y. Hu, L. Wang, J.H. Yao, H.C. Xia, J.H. Li, R. Liu, Effects of electromagnetic compound field on the escape behavior of pores in molten pool during laser cladding. Surf. Coat. Technol. **383**, 125198 (2020)
44. W. Wong, P. Vo, E. Irissou, A.N. Ryabinin, J.G. Legoux, S. Yue, Effect of particle morphology and size distribution on cold-sprayed pure titanium coatings. J. Therm. Spray Technol. **22**(7), 1140–1153 (2013)
45. T. Suhonen, T. Varis, S. Dosta, M. Torrell, J.M. Guilemany, Residual stress development in cold sprayed Al Cu and Ti coatings. Acta Mater. **61**, 6329–6337 (2013)
46. Q. Zhang, C.J. Li, X.R. Wang, Z.L. Ren, C.X. Li, G.J. Yang, Formation of NiAl intermetallic compound by cold spraying of ball-milled Ni/Al alloy powder through postannealing treatment. J. Therm. Spray Technol. **17**(5–6), 715–720 (2008)
47. M. Bray, A. Cockburn, W. O'Neill, The laser-assisted cold spray process and deposit characterisation. Surf. Coat. Technol. **203**, 2851–2857 (2009)
48. B. Li, L.J. Yang, Z.H. Li, J.H. Yao, Q.L. Zhang, Z.J. Chen, G. Dong, L. Wang, Beneficial effects of synchronous laser irradiation on the characteristics of cold-sprayed copper coatings. J. Therm. Spray Technol. **24**(5), 836–847 (2015)
49. B. Li, Y. Lin, J.H. Yao, Z.H. Li, Q.L. Zhang, X. Zhang, Influence of laser irradiation on deposition characteristics of cold sprayed Stellite 6 coatings. Opt. Laser Technol. **100**, 27–39 (2018)
50. J.H. Yao, L.J. Yang, B. Li, Z.H. Li, Characteristics and performance of hard Ni60 alloy coating produced with supersonic laser deposition technique. Mater. Des. **83**, 26–35 (2015)
51. L.J. Yang, B. Li, J.H. Yao, Z.H. Li, Effects of diamond size on the deposition characteristic and tribological behavior of diamond/Ni60 composite coating prepared by supersonic laser deposition. Diam. Relat. Mater. **58**, 139–148 (2015)
52. B. Li, J.H. Yao, Q.L. Zhang, Z.H. Li, L.J. Yang, Microstructure and tribological performance of tungsten carbide reinforced stainless steel composite coatings by supersonic laser deposition. Surf. Coat. Technol. **275**, 58–68 (2015)
53. B. Li, Y. Lin, J.H. Yao, Z.H. Li, Q.L. Zhang, Solid-state fabrication of WC_p-reinforced Stellite 6 composite coatings with supersonic laser deposition. Surf. Coat. Technol. **321**, 386–396 (2017)

Chapter 2
Magnetic Field Aided Laser Process

Abstract Conventional laser processes rely on optimizing process parameters to improve product quality and performance, thus to achieve set goals. However, the actual implementation is limited due to many factors such as time-consuming, narrow adjustable window, inter-influence between process parameters, and so on. A novel laser process controlling approach is therefore proposed, which couples a steady magnetic field to the laser system thus to assist the laser melt process, resulting in the improvement of coating surface morphology, The applied steady magnetic field suppresses the undulation of the coating surface by the dissipating effect on the molten pool dynamics. A 2D transient multi-physics numerical model, which concerns heat transfer, fluid dynamics, phase transition and magnetic field, is employed to study the suppression effect of a steady magnetic field on the molten pool during laser remelting. The dynamic shape of the laser remelting surface is explicitly simulated by an Arbitrary Lagrangian-Eulerian method (ALE). The Lorentz force generated by the steady magnetic field, the Marangoni convection formed by surface tension and the thermal buoyancy with Boussinesq approximation are all taken into account in the model. The simulation results are compared with experimental data, showing good agreement. These results demonstrate that the Lorentz force due to a steady magnetic field is a sort of drag force of the melt flow, which significantly reduces the flow velocity. The surface undulation is effectively suppressed by a steady magnetic field so that a smooth surface after laser remelting can be achieved.

2.1 Laser Surface Remelting

2.1.1 Surface Undulation Problem

Laser surface remelting technique is one of the surface modification processes [1], which allows selective heating and melting of the material surface, and modifies the surface properties by rapid melting and solidification [2–4]. The high cooling rate in the surface layer results in the formation of different microstructures from the bulk metals, leading to improved properties such as better resistance to wear, corrosion, erosion, thermal fatigue, better mechanical properties and higher hardness

J. Yao et al., *Advanced Laser Process for Surface Enhancement*, Advanced Topics in Science and Technology in China 61, https://doi.org/10.1007/978-981-15-9659-9_2

[5–8]. The modified surfaces via laser remelting are usually not post-machined, but surface undulation is always a problem because it degrades the surface quality [9]. To mitigate the undulation problem, the process parameters are optimized. However, in some cases, the process parameters not only influence the material morphology, but also affect other properties of the surface such as wear and corrosion resistance. Therefore, these parameters cannot be reasonably optimized in terms of surface morphology solely.

To solve this problem effectively, a steady magnetic field can be applied to assist the laser remelting progress, because it can suppress the undulation by dissipating effect on the molten pool dynamics. The application of magnetic field is a positive practice in many industrial processes, such as continuous casting, crystal growth, electrolysis, etc., for various purposes, for instance, grain refinement, stabilization or even destabilization of liquid surfaces, acceleration or braking of the flow of electrically-conducting fluids [10–12]. In particular, a steady magnetic field has been applied in partial penetration high power laser beam welding process [13]. The stationary simulation model and experiment were carried out to study the dissipating effect on the weld pool dynamics due to the presence of a steady magnetic field, which showed the braking of the flow velocities in the weld pool and the reduction of the evolution of spatter and melt ejections. The application of a steady magnetic field on laser alloying was studied numerically [14, 15], which demonstrated that under the influence of the applied steady magnetic field, the system of vortices was suppressed and the solute was concentrated in the shallower alloying layer.

2.1.2 Magnetic Field Application

In order to study the laser remelting process coupled with a steady magnetic field, a laser remelting experiment was conducted on the substrate material ASTM A529M, which is a low-carbon low-alloy steel, using a 2 kW LASERLINE diode laser device equipped with four NdFeB permanent magnets to create a steady magnetic field with high magnetic flux density for the molten pool [15], as shown in Fig. 2.1. The containers of permanent magnets were made of aluminum, which is a type of paramagnetic material, and were fulfilled with water to cool the magnets for maintaining their demagnetization. The transient temperature distribution was animated with a thermal imagery tool (MIKRON MCS640) and the surface morphology was measured with a fully-focused digital microscope (KEYENCE VHX-2000). The chemical composition of ASTM A529M is listed in Table 2.1. The substrate specimens were machined to long strips with dimensions of 200 × 20 × 10 mm.

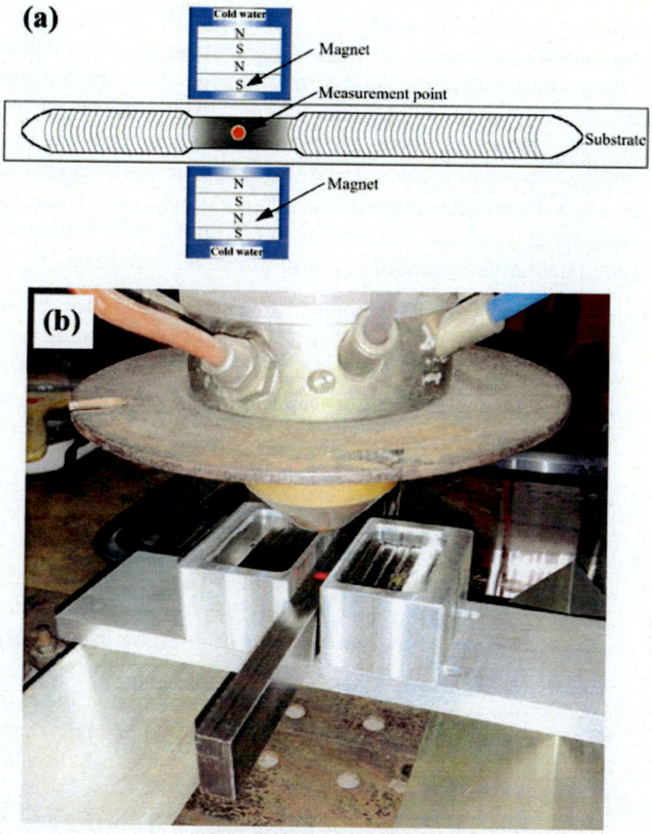

Fig. 2.1 LASERLINE diode laser system with four NdFeB permanent magnets attached: **a** sketch of specimen and magnets arrangement and **b** testbed [15]

Table 2.1 Chemical composition (wt%) of substrate material ASTM A529M

C	Mn	Si	Cr	V	Mo	Fe
0.18	0.44	0.23	<0.01	<0.01	<0.01	Bal.

2.2 Simulation of Laser Surface Remelting

2.2.1 Assumptions for Modeling

To better investigate the effects of a steady magnetic field applied on the molten pool during laser remelting, a 2D transient multi-physics numerical model, concerning heat transfer, fluid dynamics, phase transition and electromagnetic field, was created to simulate the laser remelting experiment on the ASTM A529M substrate. The model formulation involves a classical computational fluid dynamics approach that concerns fluid flow field and pressure as well as temperature. The influence of Lorentz force induced by a steady magnetic field is added in the source term of momentum equations as the volume force. The computational domain is initially composed of a rectangle of 30 × 5 mm. The surrounding gas phase is not taken into account because of the large differences in density and dynamic viscosity between the liquid metal and gas phase. The basic assumptions made on the fluid flow, temperature field and magnetic field in this simulation are summarized as follows:

(1) Laminar flow pattern is assumed. The main reason for this assumption is that the velocities outside the shear layers on the surfaces are significantly reduced and the Reynolds number inside the molten pool turns out to be far less than the turbulent critical value (10^3) [16]. Furthermore, the laminar flow assumption makes the numerical solution easier convergent and reduces the computational time [17].
(2) The material properties are temperature-dependent up to evaporation temperature.
(3) Inductive heating of the magnet system is neglected due to the strong heating of the laser source up to the evaporation temperature of the metal.
(4) The buoyancy of the molten pool is determined using Boussinesq approximation, because the density variation caused by the temperature difference is sufficiently small.
(5) The thermal conductivity is adjusted properly according to the actual situation, since the convective motion in the transverse plane (z-direction) is neglected in 2D models, which may result in the heat redistribution due to the fluid flow [14].
(6) The distribution of magnetic flux density in the active area is uniform.

2.2.2 Formulation of the Model

Mass conservation is expressed as

$$\nabla \cdot (\rho \boldsymbol{u}) = 0 \tag{2.1}$$

Momentum conservation is given as

$$\rho \frac{\partial \boldsymbol{u}}{\partial t} + \rho(\boldsymbol{u} \cdot \nabla)\boldsymbol{u} = \nabla \cdot \left[-pI + \eta\left(\nabla \boldsymbol{u} + (\nabla \boldsymbol{u})^T\right)\right] + F_{Buoyancy} + \mathrm{F}_{Darcy} + \mathrm{F}_{Lorentz} \tag{2.2}$$

where ρ is fluid density, $\boldsymbol{u}$ is fluid velocity vector, p is the pressure, I is the identity matrix, η is the dynamic viscosity and $F_{Buoyancy}$, F_{Darcy}, $F_{Lorentz}$ are source terms. The last three terms in Eq (2.2) represent the buoyancy force, Darcy term, and

Lorentz force, respectively. The buoyancy force comes from density gradients, which is related to the expansion of the liquid metal and is usually expressed using the Boussinesq approximation, as follows

$$F_{Buoyancy} = \rho(1 - \beta(T - T_m))g \tag{2.3}$$

where β is thermal expansion coefficient, g is gravivational constant, T is temperature and T_m is melting point. The role of Darcy term is to dampen the velocity at the phase change interface, therefore it becomes the velocity of the solidified phase after the transition. This term is assumed to vary with the liquid fraction in the molten pool and can be expressed as, according to the Kozeny-Carman equation [18]

$$F_{Darcy} = -A_{mush}\frac{(1 - f_L)^2}{f_L^3 + c}\boldsymbol{u} \tag{2.4}$$

where f_L is the volume fraction of the liquid phase, A_{mush} and c are arbitrary constants (A_{mush} should be a large-valued constant accounting for damping the velocity of the mushy region and c is a small constant to avoid division by zero in the solid region). The liquid fraction f_L is assumed to vary linearly with temperature in the mushy zone, which is determined by the solidus T_S and

liquidus T_L temperatures. It is defined as

$$f_L = \begin{cases} 0 & T < T_S, \\ \frac{T-T_S}{T_L-T_S} & T_S \le T \le T_L, \\ 1 & T > T_L. \end{cases} \tag{2.5}$$

The term $F_{Lorentz}$ represents the Lorentz force induced by the external steady magnetic field and the conductive fluid motion, which is described by the following equation:

$$\boldsymbol{F}_{Lorentz} = \boldsymbol{j} \times \boldsymbol{B} = \varepsilon(\boldsymbol{E} + \boldsymbol{u} \times \boldsymbol{B}) \times \boldsymbol{B} \tag{2.6}$$

where j is current density, B is magnetic flux intensity, E is external electric field (here $E = 0$), and ε is electric conductivity. The interaction of the external magnetic field with the current builds up a Lorentz force contribution, which has a component directed against the original melt velocity thus decelerating the melt. Energy conservation is given as

$$\rho C_p\frac{\partial T}{\partial t} + \rho C_p\boldsymbol{u} \cdot \nabla T = \nabla \cdot (k\nabla T) + Q_{source} \tag{2.7}$$

where Cp is heat capacity, k is thermal conductivity and Q_{source} is the heat source from laser beam. The temperature of the substrate is assumed to be ambient temperature T_0 initially. The energy distribution of laser beam is considered uniform because of the nature of diode laser. Surface tension is added on the top surface as the boundary condition, which is relevant to the surface temperature of the molten pool. In this model, surface tension is expressed by Marangoni shear stresses.

$$-\eta\frac{\partial u}{\partial z}=\frac{\partial \gamma}{\partial T}\frac{\partial T}{\partial x} \text{ and } -\eta\frac{\partial \upsilon}{\partial z}=\frac{\partial \gamma}{\partial T}\frac{\partial T}{\partial y} \tag{2.8}$$

where $\frac{\partial \gamma}{\partial T}$ is the temperature derivative of the surface tension, u and v are the fluid velocity in x and y direction respectively In order to avoid mesh distortions which may result in poor solution accuracy, an Arbitrary Lagrangian-Eulerian (ALE) method [19] is applied. In this method, the normal component of the mesh velocity u_{mesh} at the interface is equal to the normal fluid velocity.

2.2.3 *Boundary Conditions of the Model*

The boundary condition for the mesh equations on the free surface is

$$(x_t, y_t)^T \cdot \mathrm{n} = \boldsymbol{u} \cdot \mathrm{n} \tag{2.9}$$

where n is the boundary normal and $(x_t y_t)^T$ is the deformation velocity of mesh. The Winslow mesh smoothing method [20] is used in extending the moving boundary displacement throughout the domain to obtain a controlled mesh deformation. The other necessary boundary conditions are illustrated in Fig. 2.2. The physical properties of the substrate material and the laser remelting process parameters used in this simulation are given in Table 2.2 and Table 2.3, respectively.

The governing equations for mass, energy and momentum conservations including Darcy and Lorentz force sources are solved numerically using the finite element package COMSOL Multiphysics.

2.3 Characteristics of Molten Pool

2.3.1 *Temperature Distribution*

The temperature distributions in the molten pool were obtained from the experiments and simulations. The heat source with constant power density was selected to heat the specimen surface. To secure the validity of the verification, the line laser spot with a size of 3 × 8 mm was used, corresponding to the scanning speed of 4 mm/s. The same laser power and temperature monitoring point were set for both the experiment and simulation. The transient temperature variations with time computed from the simulation are reported in Fig. 2.3, compared with the experimental measurements.

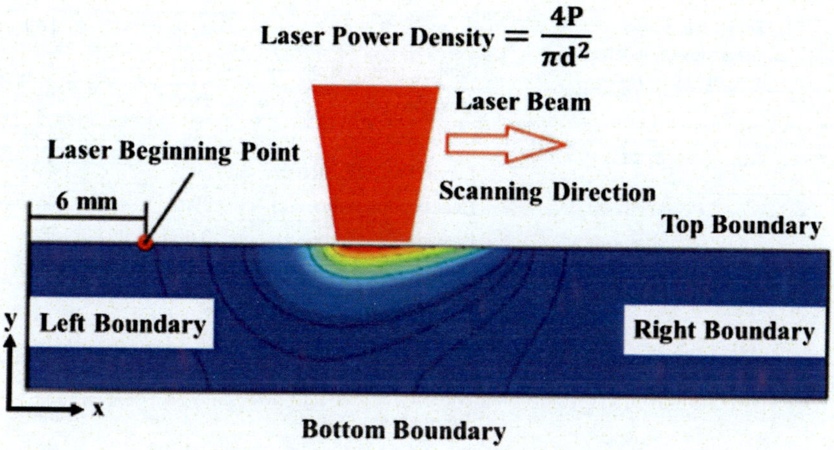

	Velocity Boundary Conditions
Top Boundary	$-\eta \frac{\partial u}{\partial y} = \frac{\partial y}{\partial T} \frac{\partial T}{\partial x}$
Left Boundary	$\mathbf{u} \cdot \mathbf{n} = 0$
Bottom Boundary	$\mathbf{u} \cdot \mathbf{n} = 0$
Right Boundary	$\mathbf{u} \cdot \mathbf{n} = 0$

	Heat Boundary Conditions
Top Boundary	$q_0 = h \cdot (T_{ext} - T)$
Left Boundary	$q_0 = h \cdot (T_{ext} - T)$
Bottom Boundary	$-\mathbf{n} \cdot (-k \nabla T) = 0$
Right Boundary	$q_0 = h \cdot (T_{ext} - T)$

Fig. 2.2 Boundary conditions of the laser remelting model [15]

Table 2.2 Physical properties of substrate material [15]

Property	Value	Unit
Melting temperature	1700	K
Mass density	7800	$kg{\cdot}m^{-3}$
Heat capacity	746	$J{\cdot}kg^{-1}{\cdot}K^{-1}$
Latent heat of fusion	30	$J{\cdot}kg^{-1}$
Heat conductivity	30	$W{\cdot}m^{-1}{\cdot}K^{-1}$
Dynamic viscosity	0.006	Pas
Surface tension coefficient	-0.52×10^{-4}	$N{\cdot}m^{-1}{\cdot}K^{-1}$

Table 2.3 Laser remelting process parameters [15]

Processing parameters	Value	Unit
Diameter of laser spot	4	mm
Laser scanning speed	6	$mm{\cdot}s^{-1}$
Laser power	1700	W

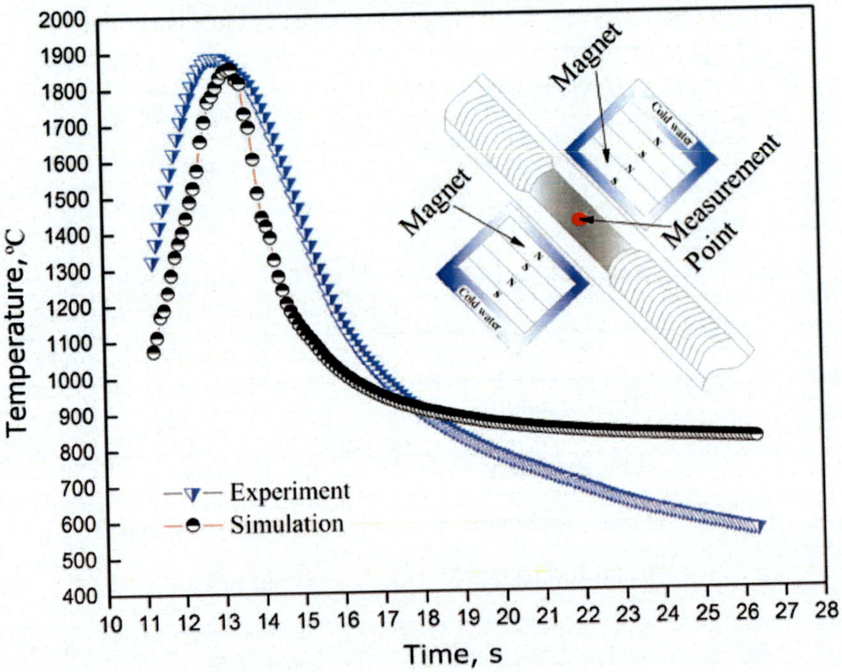

Fig. 2.3 Transient temperature variations in molten pool with time [15]

These results how good agreement, which suggests that the simulation result for the temperature distribution is acceptable.

The comparison of the temperature distributions in the molten pool with and without the steady magnetic field applied from the numerical simulation is illustrated in Fig. 2.4. The black and color contours represent the isotherm and the temperature field respectively. The magnetic flux density is 0.4 T, the laser power is 1.7 kW and

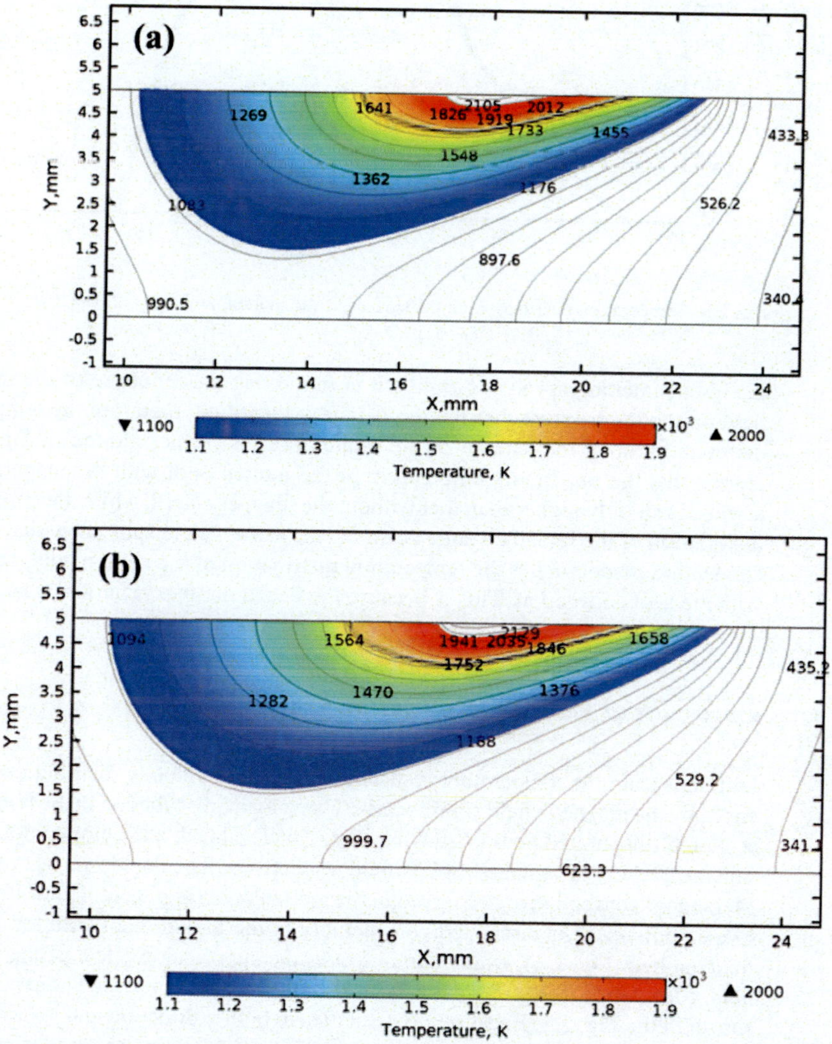

Fig. 2.4 Comparison of temperature distribution in molten pool at $t = 2.0$ s: **a** without magnetic field and **b** with magnetic field (0.4 T) [15]

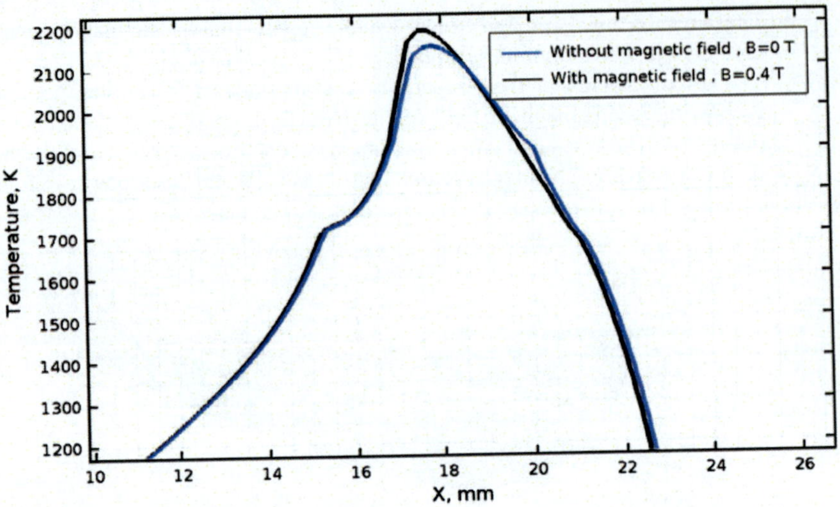

Fig. 2.5 Temperature distribution on the surface of the molten pool at $t = 2.0$ s [15]

the laser remelting speed is 6 mm/s. It is found that the influence of the magnetic field on the temperature distribution is hardly identified, therefore the temperature distribution along the surface of the molten pool is further plotted in Fig. 2.5. It is seen that the maximum temperature of the molten pool with the magnetic field applied is slightly higher than that without the magnetic field, while the temperature distribution of the remained zone is almost the same. The results show that the heat conduction predominates the temperature distribution of the molten pool, compared with the heat convection. This is because the Peclet number is far less than 1.

2.3.2 *Fluid Velocity Distribution*

Different from the temperature distribution, the fluid velocity distribution is influenced by the magnetic field significantly. The velocity distribution in the longitudinal section of the molten pool is shown in Fig. 2.6. The black and color contours represent the stream line and the velocity field respectively. The double vortices due to the Marangoni convection are observed in the molten pool (Fig. 2.6a). However, the rear Marangoni vortex is dissipated considerably in the molten pool with the magnetic field applied (Fig. 2.6), which demonstrates the suppression effect of the magnetic field. Meanwhile, the globe velocity in the molten pool is reduced considerably, as shown in the Fig. 2.6. Because the velocities in both x-direction and y–direction are orthogonal with the direction of the magnetic field, the Lorentz force is induced in the opposite direction, obtained from Eq. (2.6). Therefore, Lorentz force here is a sort of resistance force hindering the flow of the molten pool. Figure 2.7 shows the

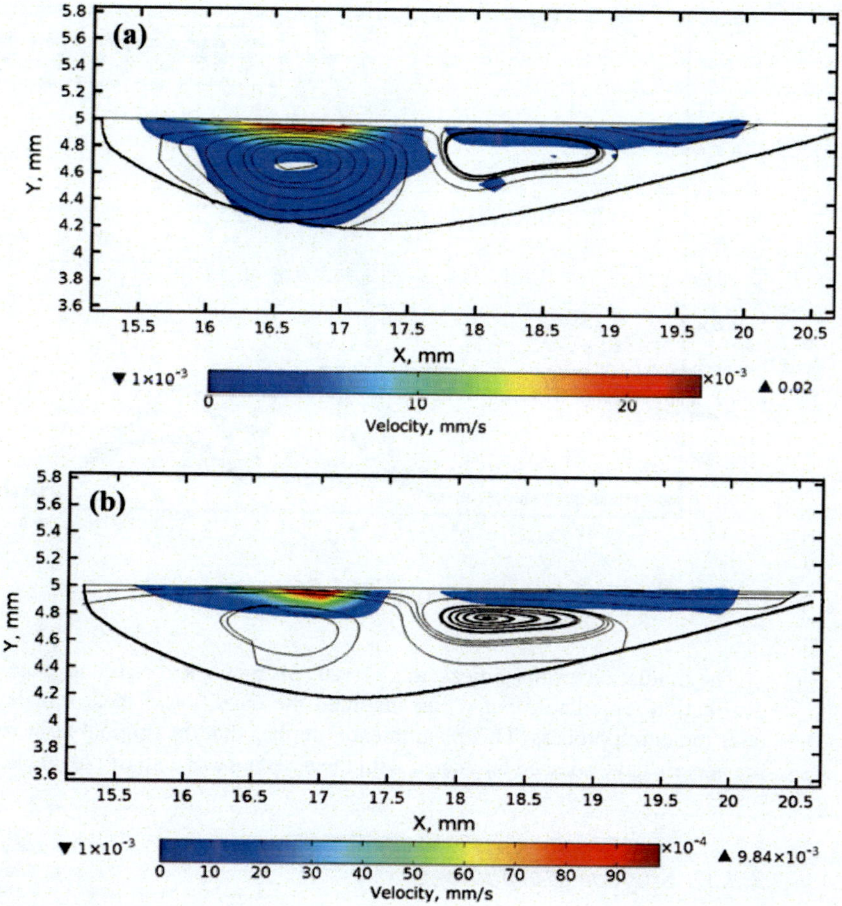

Fig. 2.6 Fluid velocity distribution in molten pool at $t = 2.0$ s: **a** without magnetic field and **b** with magnetic field (0.4 T) [15]

fluid velocity distribution at the surface of the molten pool under the laser remelting condition with and without the magnetic field applied. The double peaks depict the difference in the velocity between the two conditions because of the heat accumulation effect and Marangoni effect during the laser remelting process [21]. Owing to the damping effect, the peak velocity with the magnetic field is only about half of that without the magnetic field. There is a positive correlation in the magnitude between the Lorentz force and fluid velocity obtained from Eq. (2.6). As a result, the high velocity zone in the molten pool is shrunk more obviously. The fluid velocity of the molten pool is suppressed greatly although the magnetic flux density is only 0.4 T, as depicted in Figs. 2.6 and 2.7. It is foreseeable that with the increase of magnetic flux density, the fluid velocity of the molten pool will be reduced further.

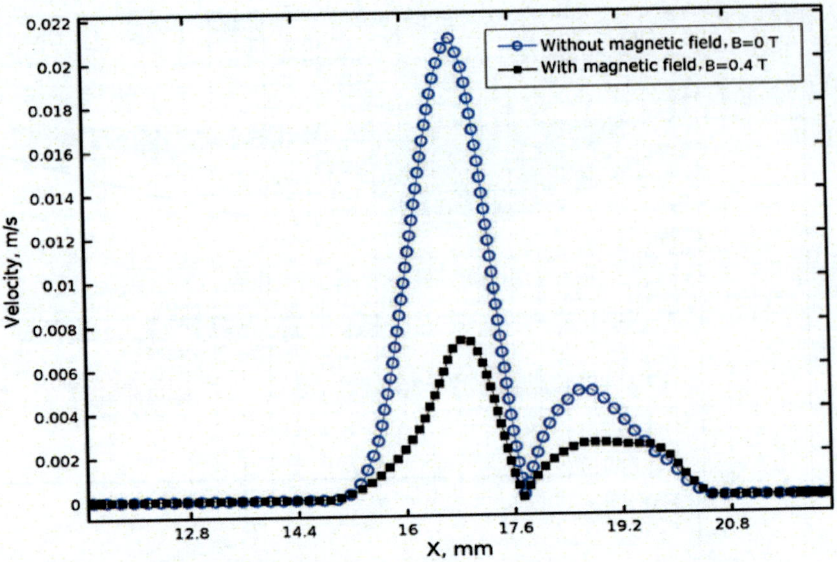

Fig. 2.7 Fluid velocity distributions on the surface of molten pool at $t = 2.0$ s [15]

The simulation results indicate that a steady magnetic field can change the velocity distribution remarkably but it can maintain the condition of heat transfer during laser remelting process. This phenomenon implies that the original laser remelting parameters have no need to change with the assistance of a steady magnetic field.

2.3.3 *Surface Morphology*

The surface morphologies of the laser remelting specimens obtained in the experimental conditions with and without the magnetic field were examined using a digital microscope. Due to the cyclic movement of the molten metal during the laser remelting, the surface undulation is unavoidable. According to the governing equations for the surface morphology, the normal component of the deformation velocity of mesh at the interface is equal to the normal fluid velocity, therefore, the surface morphology can be described precisely using the ALE method [19]. The undulation of the specimen from the laser remelting experiment is shown in Fig. 2.8a. The maximum difference between peaks and troughs of the undulation curve is measured to be about 0.1 – 0.2 mm, which is close to the numerical simulation result in Fig. 2.8b. This suggests that the numerical simulation result for the surface morphology is acceptable.

The plots in Fig. 2.9 describe the variations of the surface morphology with magnetic flux density. The normal velocity at the surface of the molten pool becomes

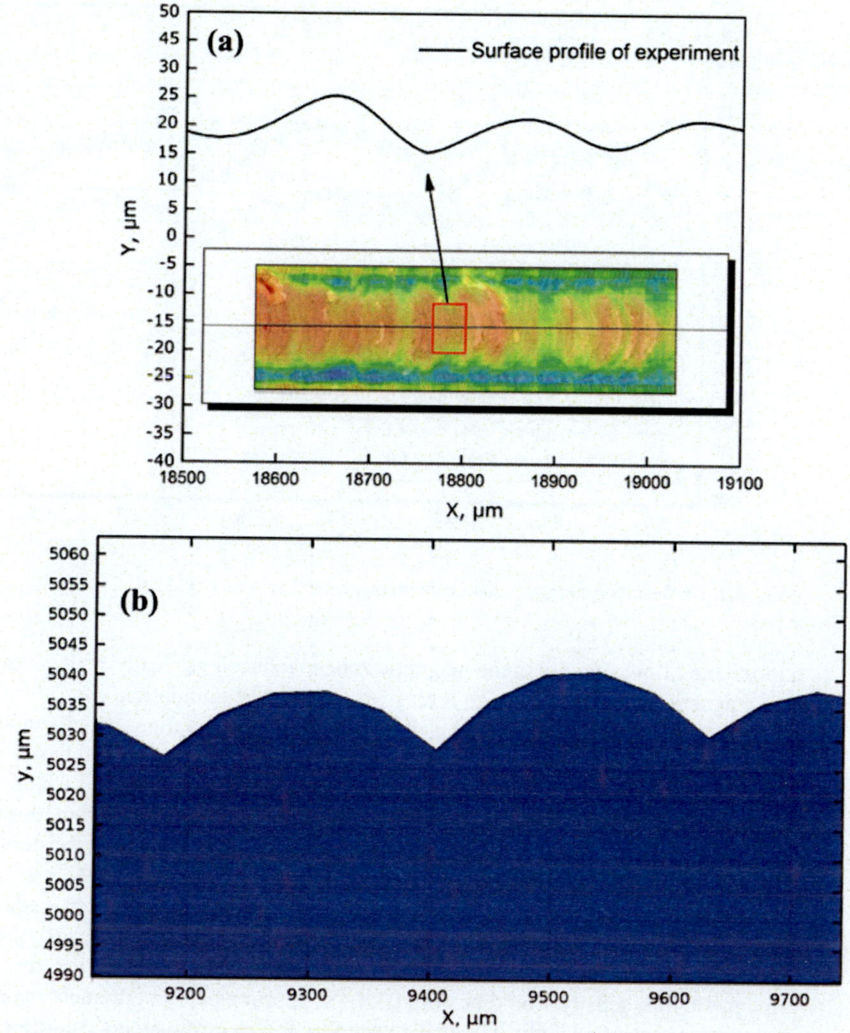

Fig. 2.8 Surface morphology of the specimen from laser remelting: **a** experimental measurement and **b** simulation prediction [15]

slower due to the suppression effect of the steady magnetic field. As a result, the undulation in the normal direction becomes smaller, which influences the overall morphology. With the increase of magnetic flux density from 0 to 0.6 T, the surface undulation is suppressed gradually. This is because the directions of Lorentz force and fluid velocity are always opposite. The magnitude of drag forces is proportional to the magnetic flux density in terms of the governing equation. The normal component

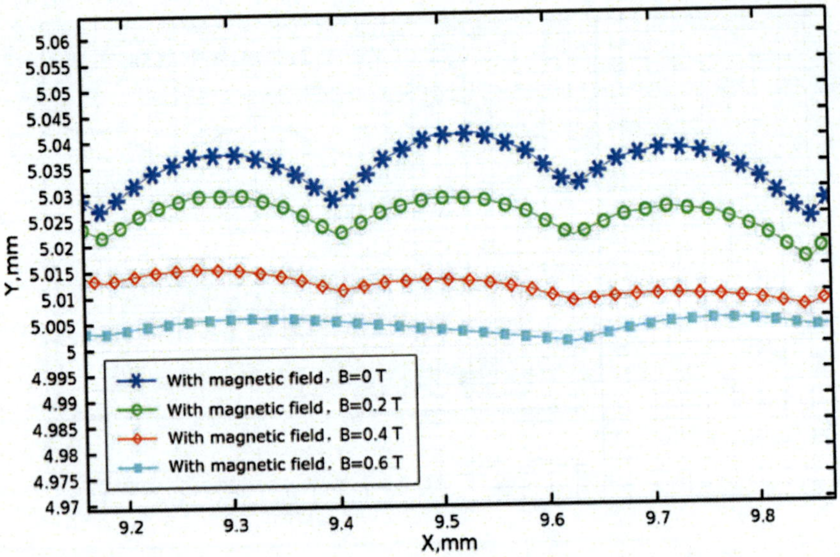

Fig. 2.9 Surface morphology variations with magnetic flux density [15]

of the melt flow velocity in the magnetic zone is reduced gradually with the increase of magnetic flux density, which represents the level of undulation.

Another concern in this study is the periodicity of undulation, which remains the same, independent of the magnitude of magnetic flux density. The simulation results show that the effect of steady magnetic field on the heat conduction is trivial, which indicates that the solidification rate of the molten pool is almost constant, despite the magnetic field. Consequently, during the continuous laser remelting, the periodicity of surface undulation remains unchanged.

According to the simulation results, in order to obtain a smooth surface, the intensity of the external magnetic flux density should be at least more than 0.4 T. When considering the assistance of magnetic field in the actual laser processing, the applied magnetic flux density would be about 0.4 – 1 T, provided by permanent magnets or electromagnets. Certainly, expensive water-cooled electromagnets can offer higher magnetic flux density, but they are only suitable for large quantity industrial process. Hence the permanent magnets that can offer about 0.4 T magnetic flux density were used in this experiment. The experimental results of surface morphology examination with the steady magnetic field are shown in Fig. 2.10. The surface undulation in the magnetic field zone is found much less than that in the magnetic-free zone, which indicates significant improvement of the surface morphology by the suppression effect of the magnetic field, although the magnetic flux density is only 0.4 T. It can be inferred that if the electromagnets with higher flux density are used, the surface undulation can be suppressed further.

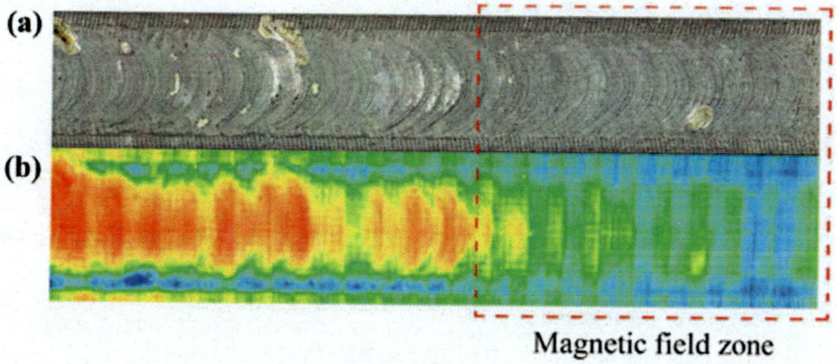

Fig. 2.10 Surface morphology of the specimen from the laser remelting experiment with the steady magnetic field (0.4 T) applied: **a** photographic morphology and **b** coloring height scanning profile showing surface undulation [15]

2.4 Summary of Magnetic Field Aided Laser Process

The suppression effect of the steady magnetic field on the molten pool during laser remelting process was investigated numerically and experimentally. The numerical model predicted the temperature and fluid velocity distributions of the molten pool as well as the surface morphology of the specimen under the laser processing conditions with and without the magnetic field applied. The simulation results were verified with the experiment measurements under 0.4 T magnetic flux density.

The main influence of the steady magnetic field on the molten pool is the damping effect that is induced by the Lorentz force in the opposite direction of the melt flow. The overall velocity of the molten metal flow and the internal double vortices are suppressed significantly by the damping effect of the magnetic field. The surface undulation of the laser remelting specimen can be reduced greatly due to the decrease of the normal melt flow velocity with the magnetic field applied. The magnetic field has little effect on the temperature distribution of the molten pool, because heat conduction pays a more significant role than heat convection in the molten pool, in terms of the small Peclet number, therefore the original laser remelting parameters can be used in the condition with the assistance of a steady magnetic field.

With the increase of magnetic flux density from 0 to 0.6 T, the surface undulation of the specimen is mitigated, but the periodicity of surface undulation remains the same, due to the same solidification cycles of the molten pool during the continuous laser remelting process, no matter if the magnetic field is applied. The application of the steady magnetic field is an effective way to minimize the surface undulation in laser remelting. Promisingly, this method can be not only applied in laser remelting process, but also used to stabilize laser cladding process.

References

1. A. Temmler, E. Willenborg, K. Wissenbach, Design surfaces by laser remelting, Phys. Procedia. **12**(A), 419–430 (2011)
2. X. Tong, M.J. Dai, Z.H. Zhang, Thermal fatigue resistance of H13 steel treated by selective laser surface melting and CrNi alloying. Appl. Surf. Sci. **271**, 373–380 (2013)
3. D. Cong, H. Zhou, Z. Ren, H. Zhang, L. Ren, C. Meng, C. Wang, Thermal fatigue resistance of hot work die steel repaired by partial laser surface remelting and alloying process. Opt. laser. Eng. **54**, 55–61 (2014)
4. P. Poza, C.J. Múnez, M.A. Garrido-Maneiro, S. Vezzù, S. Rech, A. Trentin, Mechanical properties of Inconel 625 cold-sprayed coatings after laser remelting & #x2014; Depth sensing indentation analysis. Surf. Coat. Techol. **243**, 51–57 (2014)
5. P.H. Chong, Z. Liu, P. Skeldon, P. Crouse, Characterisation and corrosion performance of laser-melted 3CR12 steel. Appl. Surf. Sci. **247**, 362–368 (2005)
6. D. Cong, H. Zhou, Z. Ren, Z. Zhang, H. Zhang, C. Meng, C. Wang, The thermal fatigue resistance of H13 steel repaired by a biomimetic laser remelting process. Mater. Des. **55**, 597–604 (2014)
7. Q.-C. Guo, H. Zhou, C.-T. Wang, W. Zhang, P.-Y. Lin, N. Sun, L. Ren, Effect of medium on friction and wear properties of compacted graphite cast iron processed by biomimetic coupling laser remelting process. Appl. Surf. Sci. **255**, 6266–6273 (2009)
8. X. Guozhi, Z. Jingxian, L. Yijun, W. Keyu, M. Xiangyin, L. Pinghua, Effect of laser remelting on corrosion behavior of plasma-sprayed Ni-coated WC coatings. Mat. Sci. Eng. A-Struct. **460–461**, 351–356 (2007)
9. E. Chikarakara, S. Naher, D. Brabazon, Spinodal decomposition in AISI 316L stainless steel via high-speed laser remelting. Appl. Surf. Sci. **302**, 318–321 (2014)
10. V.V. Avilov, A. Gumenyuk, M. Lammers, M. Rethmeier, PA position full penetration high power laser beam welding of up to 30 mm thick AlMg3 plates using electromagnetic weld pool support. Sci. Technol. Weld. Joi. **17**, 128–133 (2012)
11. Z. Tang, M. Gatzen, Influence on the dilution by laser welding of aluminum with magnetic stirring, Phys. Procedia. **5**(B), 125–137 (2010)
12. J. Zhou, H.-L. Tsai, Effects of electromagnetic force on melt flow and porosity prevention in pulsed laser keyhole welding. Int. J. Heat. Mass. Tran. **50**, 2217–2235 (2007)
13. M. Bachmann, V. Avilov, A. Gumenyuk, M. Rethmeier, About the influence of a steady magnetic field on weld pool dynamics in partial penetration high power laser beam welding of thick aluminium parts. Int. J. Heat. Mass. Tran. **60**, 309–321 (2013)
14. O. Velde, R. Gritzki, R. Grundmann, Numerical investigations of Lorentz force influenced Marangoni convection relevant to aluminum surface alloying. Int. J. Heat. Mass. Tran. **44**, 2751–2762 (2001)
15. L. Wang, J.H. Yao, Y. Hu, S.Y. Song, Suppression effect of a steady magnetic field on molten pool during laser remelting. Appl. Surf. Sci. **351**, 794–802 (2015)
16. M. Bachmann, V. Avilov, A. Gumenyuk, M. Rethmeier, Numerical simulation of full-penetration laser beam welding of thick aluminium plates with inductive support. J. Phys. D Appl. Phys. **45**(3), 035201 (2012)
17. M. Akbari, S. Saedodin, D. Toghraie, R. Shoja-Razavi, F. Kowsari, Experimental and numerical investigation of temperature distribution and melt pool geometry during pulsed laser welding of Ti6Al4V alloy. Opt. Laser Technol. **59**, 52–59 (2014)
18. A.D. Brent, V.R. Voller, K.J. Reid, Enthalpy-porosity technique for modeling convection-diffusion phase change: Application to the melting of a pure metal. Numer. Heat Trans. **13**(3), 297–318 (1988)

19. C.W. Hirt, A.A. Amsden, J.L. Cook, An arbitrary Lagrangian-Eulerian computing method for all flow speeds. J. Comput. Phys. **14**(3), 227–253 (1974)
20. M. Falsafioon, S. Arabi, R. Camarero, F. Guibault, Unstructured grid smoothing for turbomachinery applications. Int. J. Fluid Mach. Sys. **10**(4), 421–431 (2017)
21. S. Morville, M. Carin, P. Peyre, M. Gharbi, D. Carron, P. Le Masson, R. Fabbro, 2D longitudinal modeling of heat transfer and fluid flow during multilayered direct laser metal deposition process. J. Laser Appl. **24**, 1–9 (2012)

Chapter 3
Electromagnetic Field-Assisted Laser Process

Abstract During the fabrication of ceramic–metal composite coatings via laser melt injection (LMI), an issue with ceramic particle distribution within the metal matrix is often encountered. Traditionally, the successful manufacture of a coating with the desired particle distribution relies on the use of a specially designed lateral nozzle acting as a powder delivery nozzle to avoid the excessive dissolution of the reinforcement particles, thereby optimizing process parameters to achieve the required product specification. However, the practical implementation of this approach has difficulties and is also time-consuming, as the adjustable window for the lateral nozzle powder delivery system is very narrow. A novel approach to control the laser process was proposed recently, which couples an electromagnetic field to the laser system, thus assisting the laser melting process resulting in a controlled particle distribution gradient of the composite coating, as well as in controlled porosity of the coating layer. The impact of applying an electromagnetic compound field to a laser process on the produced coating is studied experimentally and numerically. In the numerical simulation, a 2D multi-physics model is created for the laser melt process coupled with the equations of heat transfer, fluid dynamics, drag force, the Lorentz force and phase transition, which incorporate the effects of the electromagnetic compound field. Experimental and simulation results obtained are in good agreement, both showing that the directional Lorentz force created by the electromagnetic compound field can change the equivalent buoyancy acting on the ceramic particles as a type of volume force, thus controlling their distribution gradient. Furthermore, with an electromagnetic compound field applied, the escape state of pores in the molten pool can be controlled effectively. With the downward Lorentz force acting, dense cladding layers with minimum porosity can be achieved successfully.

3.1 Laser Melt Injection of Composites

3.1.1 Reinforcement Particle Distribution

Metal matrix composites (MMCs) reinforced with ceramic particles can be fabricated via various processes such as casting, powder metallurgy, laser cladding, 3D printing,

J. Yao et al., *Advanced Laser Process for Surface Enhancement*, Advanced Topics in Science and Technology in China 61, https://doi.org/10.1007/978-981-15-9659-9_3

etc. However, it is the laser melt injection (LMI) method that is employed commonly to prepare an MMCs layer on a metallic substrate, which exhibits special characteristics including low particle dissolution rate, high surface performance and low cracking tendency [1–3]. Coatings processed by LMI are mainly applied to improve surface hardness and wear resistance of metallic substrates. Differently from the laser cladding process, reinforcement particles (usually ceramics) injected in the molten pool move with the melt flow and preserve solid state or micro melt state due to rapid solidification in the LMI process [4].

A graded distribution of reinforcement particles in metal matrix composites is a crucial optimization objective for the LMI process, which can be designed at a microstructural level to tailor specific materials for their functional performance in particular applications [5]. This controlled gradient in mechanical properties offers attractive challenges for the design of surfaces with resistance to contact deformation and damage [6]. In order to optimize the particle utilization ratio, machinability and gradient distribution performance of LMI layers, it is necessary to better control the distribution of reinforcement particles. Existing LMI processes commonly use a specially designed lateral nozzle as a powder delivery nozzle to avoid the excessive dissolution of reinforcement particles [7]. In this case, the key process parameters for powder injection are the injection angle with respect to the surface normal, the relative position between the powder spot and laser spot, and powder injection velocity, respectively, as they significantly influence the distribution of reinforcement particles. However, the latter is difficult and time-consuming to control through optimizing process parameters, as the adjustable window for lateral nozzle powder delivery system is very narrow [4, 8].

3.1.2 *Graded Distribution of Tungsten Carbides*

In order to control the distribution of WC particles in the composite coating during the LMI process efficiently, an external force is introduced via applying an electromagnetic compound field to the molten pool [9]. A common coaxial nozzle can still be used to replace the specially designed lateral nozzle, thus simplifying the adjusting process of the powder delivery system. The application of an electromagnetic field to alter the distribution of added elements is useful practice in laser welding and laser alloying processes. Effects of electromagnetic stirring on element distribution in laser welding have been investigated by numerical and experimental methods. It was shown that the change in filler material distribution resulted from the modulation of melt flow due to periodically induced electromagnetic volume forces [10]. Frequency is the main parameter determining the spatial distribution of elements, whereas magnetic flux density is the principal factor shaping the overall scale of magnetic manipulation [11].

3.2 Laser Melt Injection Process

3.2.1 Coating Materials

The assistive role of an external electromagnetic field in the laser melt injection process was investigated experimentally by creating WC reinforced composite coating on AISI 316L austenitic stainless steel substrate [9] with paramagnetic properties. The specimen with dimensions of 200 × 20 × 10 mm and a chemical composition of AISI 316L is detailed in Table 3.1. The WC particles had a spherical shape and were sized 75 – 150 μm for good tracing performance in the melt flow, as shown in Fig. 3.1.

Table 3.1 Chemical composition (wt%) of AISI 316

C	Si	Mn	P	S	Ni	Cr	Mo	Fe
0.02	0.55	1.55	<0.03	<0.03	10.0	16.5	2.08	Bal.

Fig. 3.1 SEM morphology of WC particles [9]

3.2.2 Electromagnetic Compound Field

Figure 3.2 illustrates the setup for the LMI process coupled with an electromagnetic compound field. Electromagnets were used to provide a steady magnetic field with a maximum magnetic flux density of 2.0 T; large capacity lead-acid batteries (2 V, 500 Ah) provided a steady electric field with high current for the molten pool. The size of magnet poles was 80 × 20 mm. The specimen was placed in the central zone of the magnetic gap where the distribution of magnetic flux density was uniform. The relative magnetic permeability ($\mu_r = 1$) of AISI 316L is almost the same as that of air. Magnetic flux density in the molten pool zone was approximately 0.4 – 1.2 T, whereas the average current density was approximately 5 A/mm^2.

The LMI experiment was conducted using a 2 kW Laserline diode laser beam with custom made powder feeder. Argon was used as shielding gas to reduce the oxidation of the specimen substrate and WC particles. Laser beam diameter, optimal laser power, scanning speed and powder feeding ratio in this experiment were 4 mm, 1.7 kW, 4.25 mm/s and 15 g/min, respectively. The distribution of WC particles in the molten pool were examined using a scanning electron microscope (SEM) (Carl Zeiss SIGMA HV-01–043). A common coaxial nozzle was used with both an external steady electric field and a steady magnetic field applied synchronously in the molten pool during the LMI process. The electromagnetic synergistic effect generated the Lorentz force, mainly a type of directional volume force in the molten pool, which is similar to gravity. This force can function as an additional volume force acting on the melt flow with variable direction. Consequently, locations of WC particles trapped in the molten pool can be altered or controlled without changing LMI process parameters.

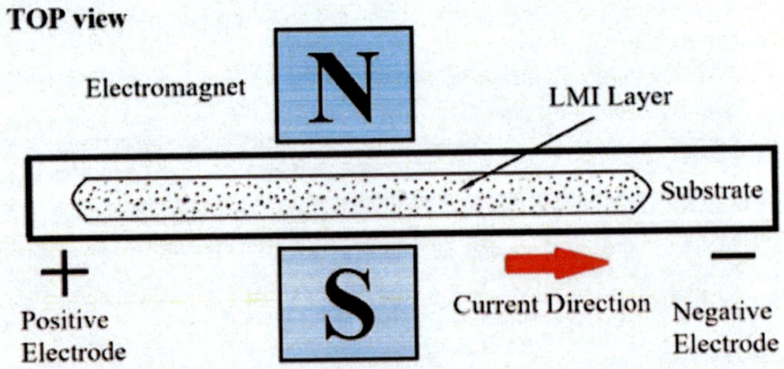

Fig. 3.2 Schematic diagram of the setup of LMI process coupled with electromagnetic compound field [9]

3.3 Simulation of Electromagnetic Field-Assisted Laser Melt Injection

3.3.1 *Assumptions for Model Formulation*

To better investigate the effects of an electromagnetic compound field on the molten pool in the LMI process, and of the influencing mechanisms of the electromagnetic compound field in particle distribution within the LMI composite coating, a numerical simulation of the LMI experiment for fabricating WC/AISI 316L stainless steel composite was performed by creating a 2D transient multi-physics model [9]. An approach involving classical computational fluid dynamics was applied to calculate fluid flow field and pressure, as well as temperature in the molten pool. The movement of injected particles was computed using the Lagrangian approach with fluid-particle coupling [12]. The Lorentz force is included in the source term of momentum equations. The 2D finite element model is composed of a rectangle of 30 $\times$ 5 mm. Necessary assumptions for the simulation model are made as follows.

(1) The fluid has laminar flow, and is incompressible Newtonian fluid [13, 14]. Thermal conductivity, electrical conductivity, and heat capacity of the specimen material are temperature dependent. Other material properties are constant.
(2) Radiation loss and Joule heat induced by the current are negligible due to short action time and the high performance heating of the laser source [15].
(3) The Boussinesq approximation is used to take buoyancy in the molten pool into account.
(4) The distribution of magnetic flux density in the active area is uniform.
(5) For the conditions of the experimental setup, the distance between the nozzle and the substrate sample is at defocusing state (focus + 3 mm). The resultant covering area of the powder stream is slightly larger than the area of laser beam. Powder distribution at the surface of the molten pool is nearly uniform in the zone of laser beam [16]. In order to simplify the computation of the model, typical injection parameters are assumed as follows: the direction of particle injection is perpendicular to the molten pool surface, and particles are uniformly distributed in the powder stream.
(6) In the LMI process, injected particles maintain a non-melting state, with the shape of injected particles being a regular sphere. Therefore, the drag force (Stoke's force) acting on particles is only related to the velocity of melt flow [4].
(7) Shear stress induced by the carrier gas on the molten pool surface is negligible due to the differences of magnitude in density and dynamic viscosity between the liquid metal and the gas phase [17].
(8) Heat and convection in the molten pool are not influenced by the injection of particles.
(9) Under the LMI conditions, calculation results based on literature about pushing/engulfment transition show that interfacial force acting on particles

can be neglected [15]. In addition, interfacial force in laser melt injection is usually also eliminated in literature [13].

3.3.2 Theories for Model Formulation

Mass conservation is expressed as

$$\nabla \cdot (\rho \boldsymbol{u}) = 0. \tag{3.1}$$

Momentum conservation is given as.

$$\rho \frac{\partial \boldsymbol{u}}{\partial t} + \rho(\boldsymbol{u} \cdot \nabla)\boldsymbol{u} = \nabla \cdot [-pI + \eta(\nabla \boldsymbol{u} + (\nabla \boldsymbol{u})^T)] + F_{Buoyancy} + \mathrm{F}_{Darcy} + \mathrm{F}_{Lorenz} \tag{3.2}$$

where ρ is fluid density, $\boldsymbol{u}$ is fluid velocity vector, p is pressure, I is identity matrix, η is dynamic viscosity and $F_{Buoyancy}$, F_{Darcy}, $F_{Lorentz}$ are source terms. The last three terms in Eq. (3.2) represent Buoyancy force, Darcy term and Lorentz force, respectively. Buoyancy force is generated by density gradients, is related to the expansion of liquid metal, and is usually expressed using the Boussinesq approximation as follows:

$$F_{Buoyancy} = \rho(1 - \beta(T - T_m))g \tag{3.3}$$

where β is thermal expansion coefficient, g is gravivational constant, T is temperature and T_m is melting point. The role of Darcy term is to dampen velocity at the phase change interface, therefore it becomes the velocity of the solidified phase after transition. This term is assumed to vary with the liquid fraction in the molten pool, and can be expressed according to the Kozeny-Carman equation [18] as

$$F_{Darcy} = -A_{mush} \frac{(1 - f_L)^2}{f_L^3 + c} \boldsymbol{u} \tag{3.4}$$

where f_L is the volume fraction of the liquid phase, A_{mush} and c are arbitrary constants (A_{mush} should be a constant of a large value accounting for damping the velocity of the mushy region and c is a small constant to avoid division by zero in the solid region). The liquid fraction f_L is assumed to vary linearly with temperature in the mushy zone, which is determined by the solidus T_S and liquidus T_L temperatures. It is defined as

$$f_L = \begin{cases} 0 & T < T_S, \\ \frac{T - T_S}{T_L - T_S} & T_S \le T \le T_L, \\ 1 & T > T_L. \end{cases} \tag{3.5}$$

The constants A_{mush} and c in Eq. (3.4) are 6×10^3 and 0.001, respectively in this model. $T_S = 1670$ K and $T_L = 1730$ K in Eq. (3.5). The Lorentz force induced by the steady external magnetic field, steady electric field and conductive fluid flow motion are described by the following equation:

$$\boldsymbol{F}_{Lorentz} = \boldsymbol{j} \times \boldsymbol{B} = \varepsilon(\boldsymbol{E} + \boldsymbol{u} \times \boldsymbol{B}) \times \mathrm{B} = \varepsilon(\boldsymbol{E} \times \boldsymbol{B}) + \varepsilon(\boldsymbol{u} \times \boldsymbol{B} \times \boldsymbol{B}). \quad (3.6)$$

Here, $E \neq 0$. The interaction between the external magnetic field and the external electric field generates a directional Lorentz force contribution $\varepsilon(\boldsymbol{E} \times \boldsymbol{B})$. Simultaneously, the motion of conductive fluid flow in the external magnetic field generates the induced electric current. This induced current in the external magnetic field forms the induced Lorentz force ($(\varepsilon \boldsymbol{u} \times \boldsymbol{B} \times \boldsymbol{B})$), which has a component directed against the original melt velocity thus decelerating the melt flow. Energy conservation is given as.

$$\rho C_p \frac{\partial T}{\partial t} + \rho C_p \boldsymbol{u} \cdot \nabla T = \nabla \cdot (k \nabla T). \quad (3.7)$$

Particle movement in the molten pool follows Newton's second law:

$$\frac{d(m_p v)}{dt} = F_{Darcy} + F_{Gravity} + F_{Buoyancy} \quad (3.8)$$

$$F_{Darcy} = m_p \left(\frac{1}{\tau_p}\right)(u - v) \quad (3.9)$$

$$F_{Gravity} = m_p g \quad (3.10)$$

$$F_{Buoyancy} = -V_p \rho g \quad (3.11)$$

where F_{Darcy} is the drag force acting on the particle, $F_{Gravity}$ is the gravity of the particle, $F_{buoyancy}$ is the buoyancy force acting on the particle, m_p is particle mass, u is fluid velocity, v is particle velocity, ρ is fluid density and V_p is the volume of a single particle. The relative Reynolds number varies from 1 to 100, and the drag coefficient C_D based on Schiller-Naumann Drag Mode is calculated as follows [14]:

$$Re_r = \frac{\rho|u - v|d_p}{\eta} \quad (3.12)$$

$$C_D = \frac{24}{Re_r}\left(1 + 0.15 {Re_r}^{0.637}\right) \quad (3.13)$$

$$\tau_p = \frac{4\rho_p d_p^2}{3\eta C_D Re_r} \quad (3.14)$$

where Re_r is the relative Reynolds number, d_p is the particle diameter and ρ_p is the particle density.

The initial substrate temperature is assumed to be ambient temperature T_0. The energy distribution of the laser beam is considered uniform because of the nature of the laser diode. Surface tension is added to the top surface as the boundary condition, which is relevant to the molten pool surface temperature. In this simulation, surface tension acts both in the normal direction and the tangential direction, since free surface can deform, which is defined as

$$\text{Normal direction: } \sigma_n = -\kappa\gamma \cdot n \tag{3.15}$$

$$\text{Tangential direction: } \sigma_m = \frac{\partial\gamma}{\partial T}\nabla T \cdot m \tag{3.16}$$

where n is the normal vector, m is the tangential vector, κ is surface curvature. The laser energy Q_{source} is added by moving heat source as the boundary heat flux to the top of the solution domain, which is defined as

$$Q_{source}(x, t) = \begin{cases} \frac{\alpha cos(\theta) P}{\pi r_l{}^2} & |x - V_s t| \le r_l \\ 0 & |x - V_s t| > r_l \end{cases} \tag{3.17}$$

where α is material absorptivity, θ is the angle between the surface normal and the laser direction, that is, incidence angle, P is laser power, r_l is laser beam radius and V_S is laser beam scanning speed. The injection velocity of WC particles is calculated as follows [19]

$$P_u = \frac{3D^2 V_0{}^2}{32}\frac{cos\theta}{H} \tag{3.18}$$

where D is nozzle hydraulic diameter, V_0 is carrier gas velocity at the nozzle, χ is kinematic viscosity of the delivering gas, θ is particle divergence angle, and H is the distance between nozzle and substrate sample. Particle injection velocity was calculated to be approximately 1.1 m/s, while the powder spot center coincides with the laser spot center.

3.3.3 Boundary Conditions and the Solution of the Model

The ALE method [20] was also utilized in this simulation to track the motion of the free interface with the Comsol Multiphysics® software using the Winslow smoothing method [20]. The boundary condition for controlling the free surface is defined as:

$$V_m = \boldsymbol{u} \cdot n \tag{3.19}$$

where V_m is normal mesh deformation velocity. Other boundary conditions required are illustrated in Fig. 3.3.

The coupled equations are solved using the mesh consisting of 6006 quadrilateral elements with a maximum element size of 0.13 mm in the surface domain. The solution is performed with the direct solver PARDISO [20] associated with the generalized-α temporal method. Parameters for the nozzle and carrier gas are reported in Table 3.2. The physical properties of the molten pool and WC particles are summarized in Table 3.3 [21]. Temperature-dependent properties of the AISI 316L steel are depicted in Fig. 3.4 [22].

3.4 Molten Pool Characteristics

3.4.1 Fluid Flow Velocity Distribution

The fluid velocity distribution in the longitudinal section of the molten pool is shown in Fig. 3.5, where *MN* represents the direction of the Lorentz force. When $MN = 1$, this direction is opposite to gravity force. When $MN = -1$, it coincides with that of gravity force. The red arrows indicate the motion direction of fluid flow; the length of arrows corresponds to the magnitude of the motion velocity of fluid flow and the green line denotes the molten pool boundary. The process is characterized by a magnetic flux density of 0.8 T, current density of 5 A/mm^2, laser power of 1.7 kW, scanning speed of 4.25 mm/s and powder feeding rate of 15 g/min. Due to the Marangoni convection, double vortices are observed in the molten pool and fluid velocity in the near surface of the molten pool is much higher than that in the rest zone, as shown in Fig. 3.5a. However, global fluid velocity is reduced considerably in the molten pool with the electromagnetic compound field applied, as shown in Figs. 3.5b–d. With the effect of an electromagnetic field, there are two types of the Lorentz force generated in the molten pool simultaneously: one is the directional Lorentz force and the other is the induced Lorentz force, as expressed in Eq. (3.6). The main mechanism for fluid velocity behavior in Fig. 3.5b is the contribution of the induced Lorentz force, which supports the suppression effect of a steady magnetic field [23]. Compared with the effect of the induced Lorentz force, the effect of the directional Lorentz force on the velocity distribution field is limited. In this case, the suppression effect of the magnetic field predominates fluid velocity variation. Since the directional Lorentz force induced by the electromagnetic compound field as a type of uniform volume force acting on the entire molten pool, it can alter the equivalent gravity force, but hardly changes fluid flow velocity.

Therefore, it is necessary to analyse different effects between the induced Lorentz force due to fluid motion in a steady magnetic field, and the directional Lorentz force induced by the electromagnetic compound field. The surface fluid velocity of the

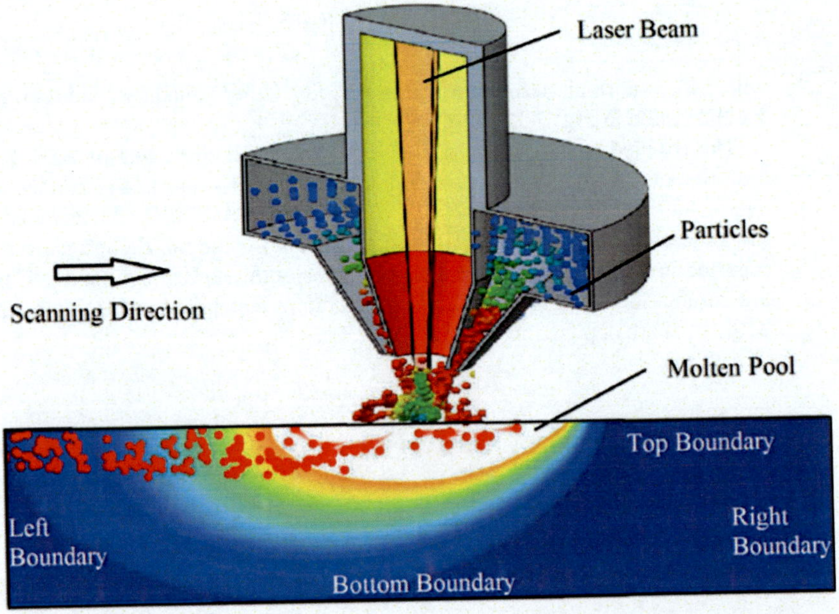

	Velocity Boundary Conditions
Top Boundary	$-\eta u_{i,k} = \frac{\partial\gamma}{\partial T}\frac{\partial T}{\partial x_i}(i=1,2;k=3)$
Left Boundary	$\mathbf{u}\cdot\mathbf{n}=0$
Bottom Boundary	$\mathbf{u}\cdot\mathbf{n}=0$
Right Boundary	$\mathbf{u}\cdot\mathbf{n}=0$
	Heat Boundary Conditions
Top Boundary	$q_0 = h\cdot(T_{ext}-T)$
Left Boundary	$q_0 = h\cdot(T_{ext}-T)$
Bottom Boundary	$-\mathbf{n}\cdot(-k\nabla T)=0$
Right Boundary	$q_0 = h\cdot(T_{ext}-T)$

Fig. 3.3 Boundary conditions of the numerical model [9]

Table 3.2 Nozzle and carrier gas parameters [9]

Parameter	D	V_0	χ	θ	H
Value	1 mm	2 m/s	15.7×10^{-6} m^2/s	8°	22 mm

Table 3.3 Physical properties of molten pool and particles [9]

Property	Symbol	Value	Unit
Melting point	T_m	1700	K
Fluid density	ρ	7850	kg·m^{-3}
Dynamic viscosity	η	0.006	Pa·s
Thermal expansion coefficient	β	5×10^{-5}	K^{-1}
Heat convection coefficient	h	20	W·m^{-2}·K^{-1}
Surface tension coefficient	γ	1.76	N/m
Temperature derivative of surface tension	$\frac{\partial \gamma}{\partial T}$	-5.2×10^{-4}	N·m^{-1}·K^{-1}
Particle density	ρ_p	15,600	kg·m^{-3}
Particle diameter	d_p	80	µm
Absorptivity	α	0.4	–
Laser beam radius	r_1	2	mm

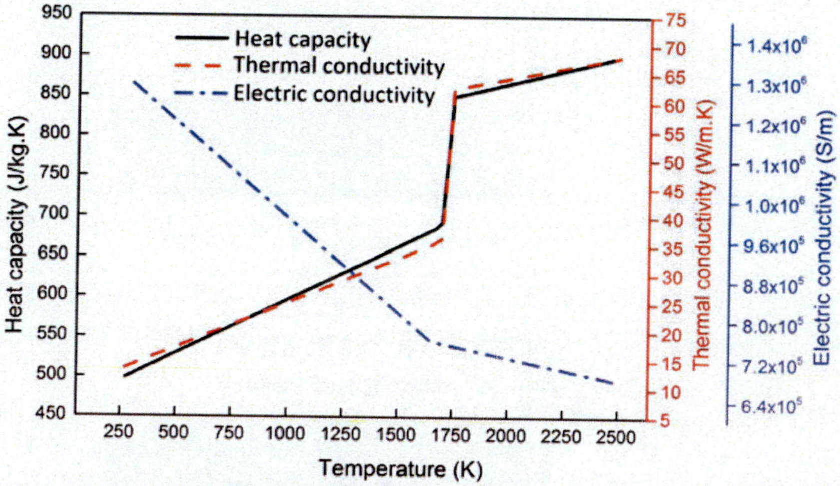

Fig. 3.4 Temperature-dependent properties of AISI 316L steel [9]

molten pool with a steady-only magnetic field is plotted in Fig. 3.6. A double-peak pattern can be observed in surface fluid velocity along the molten pool. During the laser scanning process, thermal gradients are opposite in sign on each side of the peak temperature location, which leads to the fluid flow velocity being equal to zero

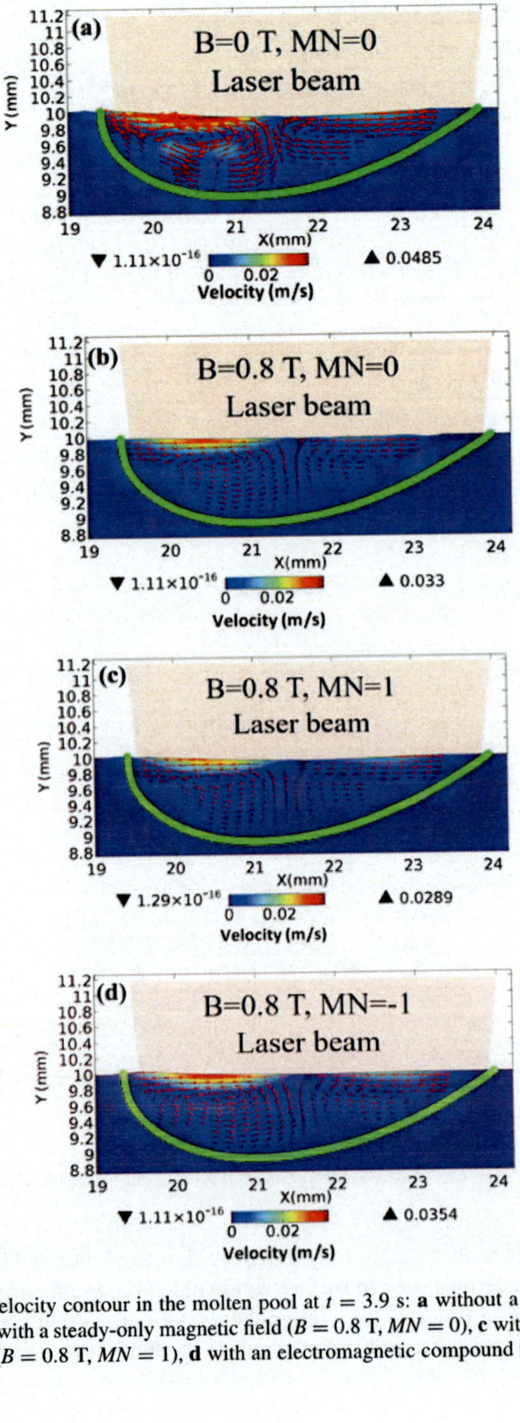

Fig. 3.5 Fluid velocity contour in the molten pool at $t = 3.9$ s: **a** without a magnetic field ($B = 0$ T, $MN = 0$), **b** with a steady-only magnetic field ($B = 0.8$ T, $MN = 0$), **c** with an electromagnetic compound field ($B = 0.8$ T, $MN = 1$), **d** with an electromagnetic compound field ($B = 0.8$ T, $MN = -1$) [9]

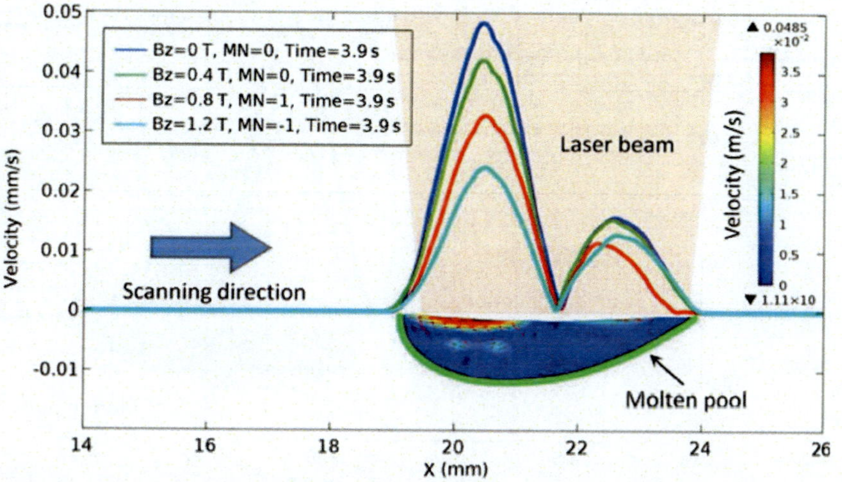

Fig. 3.6 Fluid flow velocity distribution on the molten pool surface in longitudinal direction for the condition with different magnetic flux densities [9]

at the point due to the Marangoni effect. Thermal gradients peak at the edge of the laser beam, which explains the velocity peaks observed [16]. With the increase of the magnetic flux density from 0 to 1.2 T, maximum velocity at the surface of the molten pool is gradually reduced from 0.048 to 0.024 m/s, as illustrated in Fig. 3.6. Once magnetic flux density reaches 1.2 T, peak velocity is only about half of that in the condition without a magnetic field applied. This is because the direction of the induced Lorentz force due to a steady magnetic field is always opposite to the direction of the fluid flow defined by Eq. (3.6). These results confirm the strong damping effect of a steady magnetic field in line with previous studies [15].

To further investigate the relationship between the velocity distribution and the directional Lorentz force, surface fluid velocity of the molten pool with the electromagnetic compound field present is plotted in Fig. 3.7. The magnetic flux density of the steady magnetic field is set to 0.8 T. The direction of the directional Lorentz force is kept unchanged and is opposite to gravity force. Regardless of whether a steady magnetic field or an electromagnetic compound field is applied, fluid velocity at the surface of the molten pool decreases greatly due to the suppression effect of the steady magnetic field. However, with the electromagnetic field coupled, maximum fluid velocity at the surface of the molten pool changes slightly with the variation of the direction of the directional Lorentz force. Maximum fluid velocity at the molten pool surface is higher with $MN = -1$ than with $MN = 1$ or $MN = 0$. When $MN = 1$, maximum surface fluid velocity of the molten pool is even lower than that with only the steady magnetic field applied.

Surface tension is the main driving force for fluid motion in the molten pool [16], but the secondary factor that can influence fluid motion due to the non-uniformity of temperature distribution in the molten pool is natural convection, which is expressed

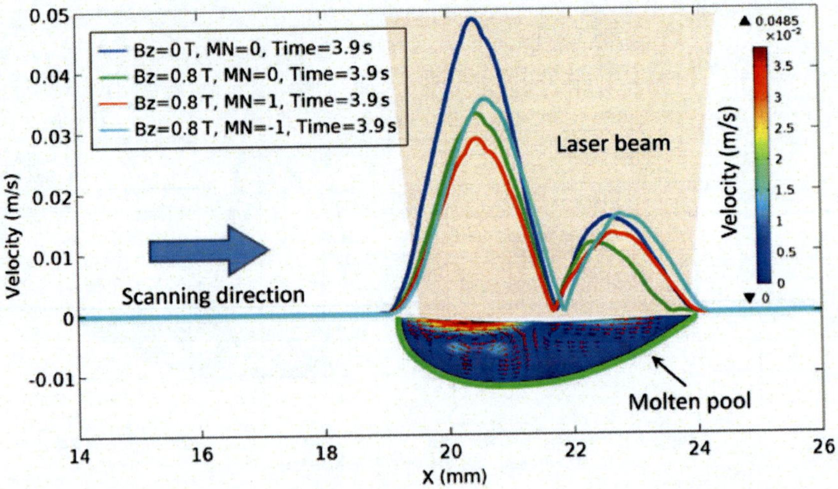

Fig. 3.7 Velocity distributions on the molten pool surface in longitudinal direction, for the condition with different directions of the Lorentz force [9]

by the Boussinesq approximation Eq. (3.3). When the directional Lorentz force, as a type of volume force, acts in the same direction as gravity force, the equivalent gravity accelerant can be assumed to increase. The resultant natural convection is enhanced accordingly, which accelerates fluid flow by a certain amount. When the direction of the directional Lorentz force is opposite to gravity force, the effect of natural convection is reduced and the fluid flow velocity is even further suppressed. From simulation results as shown in Figs. 3.5 and 3.7, it is evident that the velocity distribution of the molten pool is associated with the direction of the directional Lorentz force, but its influence is less than that of the induced Lorentz force caused by a steady magnetic field.

3.4.2 Temperature Distribution

Prior investigation results that are relevant had verified that, temperature distribution in the molten pool hardly changes under the effect of a steady magnetic field during the laser remelting process [15]. In order to compare temperature distributions in the molten pool with and without an electromagnetic compound field applied, temperature distribution on the molten pool surface in longitudinal direction is plotted in Fig. 3.8. It is shown that when the steady magnetic field or the electromagnetic compound field is applied, temperature in the central zone of the molten pool surface is only slightly higher than that without any external fields applied. The maximum rising range of temperature is merely about 50 K. Temperature values for the edge

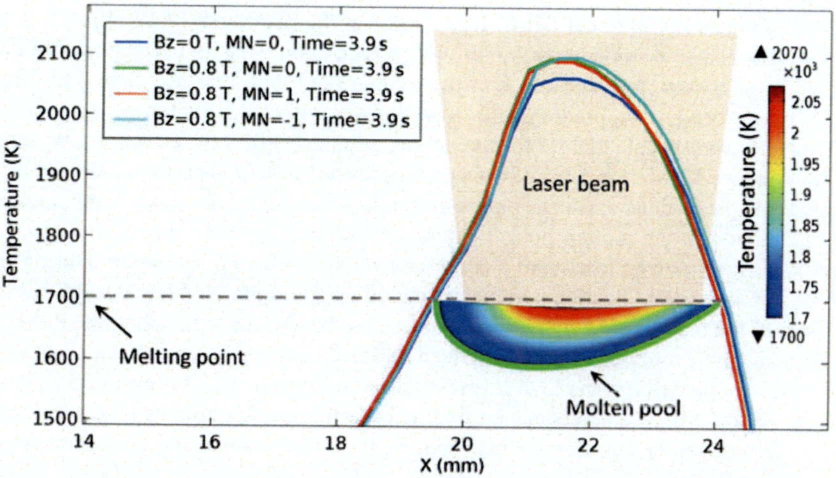

Fig. 3.8 Temperature distribution on the molten pool surface in longitudinal direction for the condition with different directions of the Lorentz force [9]

of the molten pool and those for unmelted zones are almost the same. Temperature distribution in the molten pool is hardly influenced by the direction of the directional Lorentz force, and is also barely influenced by the electromagnetic compound field. It is inferred that with the present process parameters, heat conduction predominates temperature distribution in the molten pool compared with heat convection, which results in the change of fluid velocity being insensitive to molten pool temperature distribution.

In order to verify the above speculation, the Peclet number (Pe) as a dimensionless number is introduced to show the ratio of heat convection to heat conduction. Figure 3.9 illustrates the numerical values for Pe distribution in the molten pool

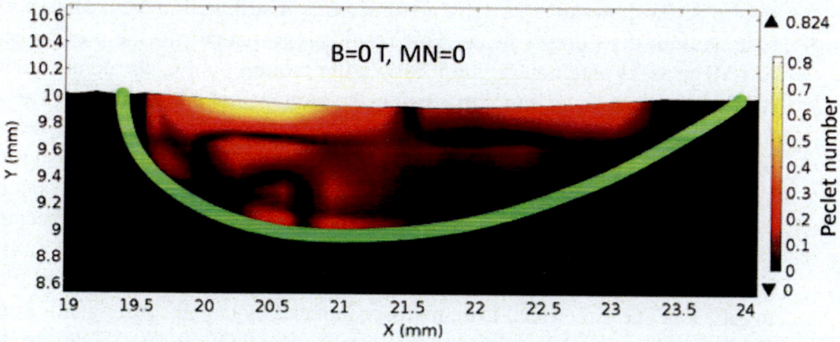

Fig. 3.9 Peclet number distribution within the molten pool [9]

without any external fields. It is clear that the maximum of Pe appears at the rear surface of the molten pool with a magnitude of only 0.824, where peak fluid velocity also occurs. In other zones of the molten pool, Pe is far less than one. Therefore, heat conduction predominates heat transfer within the molten pool in those zones. It is foreseeable that Pe for the molten pool, as well as the effect of heat convection induced mainly by fluid flow will be reduced further with the addition of a steady magnetic field or electromagnetic field. It can be inferred that temperature distribution in the molten pool is mainly determined by laser process parameters such as laser power, laser spot size, laser scanning velocity, and so on. During the LMI process, solidification parameters including the depth, width and solidification time of the molten pool are almost the same for conditions with and without the external magnetic or compound field. Meanwhile, the degree of decomposition of reinforcement particles in the LMI layer remain nearly the same. This means that the original laser LMI parameters can be retained whether the electromagnetic compound field is applied or not, eliminating the need for a time-consuming adjustment of process parameters.

3.4.3 Particle Distribution

The exact locations of trapped reinforcement particles are predicted using a simulation model with conditions of a magnetic flux density of 0.8 T, current density of 5 A/mm^2, scanning speed of 4.25 mm/s, laser power of 1.7 kW and powder feeding rate of 15 g/min, respectively. Results are presented in Fig. 3.10 for particle distributions with different directions of the Lorentz force in the longitudinal sections of the LMI layers. Red dots represent the particles injected, tail orientation indicates the direction of particle motion, whereas tail length measures the magnitude of particle motion velocity.

It was found that under the same LMI process parameters, the number of injected particles, molten pool depth and solidification time all remain unchanged. Figure 3.10c shows that particles are uniformly distributed in the molten pool in the case without the Lorentz force. This result suggests that, for the present condition of LMI process parameters, fluid flow in the molten pool is stirred adequately and reinforcement particles can be trapped in the molten pool potentially anywhere due to the drag force acting in the melt flow. It is interesting to notice that with the presence of the Lorentz force, reinforcement particle distributions are changed significantly. When the direction of the Lorentz force is upward, most particles are concentrated in the lower region of the LMI layer, as shown in Fig. 3.10a. This is because fluid velocity in the near-surface zone (Marangoni convection zone) of the molten pool is damped greatly with the effect of electromagnetic compound field, as illustrated in Fig. 3.7. The directional Lorentz force induced by the electromagnetic compound field is a type of volume force that is similar to gravity force. When the direction of the Lorentz force is opposite to gravity force, the equivalent gravity force acting on the fluid of molten pool is reduced. The resultant buoyancy acting on particles is

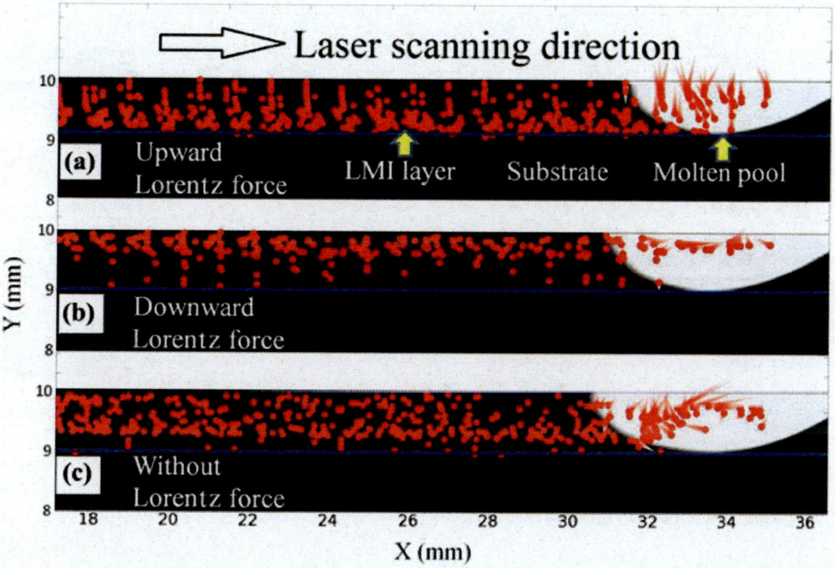

Fig. 3.10 Simulation prediction of WC particle distribution in longitudinal section of the LMI layer under different conditions: **a** the Lorentz force is upward, **b** the Lorentz force is downward and **c** without the Lorentz force [9]

decreased accordingly. Hence, the sinking velocity of particles is increased. Therefore, when particles are injected into the molten pool, they can quickly penetrate the Marangoni convection zone at the surface of the molten pool and sink to the lower region of the molten pool easily before it solidifies. With the effect of the downward Lorentz force of the direction same to gravity force, buoyancy acting on particles increases. When particles are injected into the molten pool, their sinking velocity is reduced rapidly. As a result, particles can hardly penetrate the near-surface zone, thus are easier dragged to the edge of the molten pool, even if the Marangoni convection zone where the molten pool is prone to solidify is highly suppressed, as shown in Fig. 3.10b. Consequently, most particles get trapped in the upper region of the molten pool. It is noted that the final trapped location of a particle is also related to the synchronously injecting position. In this model, the distribution of particles on the surface of the molten pool before injection into the molten pool is assumed uniform. In addition, simulation results show that the electromagnetic compound field hardly influences temperature distribution in the molten pool, therefore, this field would not change the solidification rate and the depth of the molten pool. Accordingly, the difference in the overall distribution of injected particles under the same modeling boundary conditions during the LMI process is certainly caused by the effect of the directional Lorentz force.

The distribution of WC particles in the coating layer predicted from the simulation was compared with that from the LMI experiment. Figure 3.11 shows distribution in

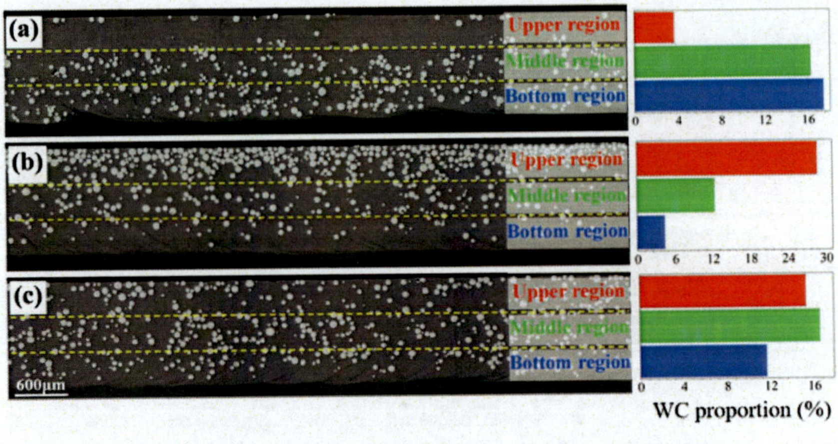

Fig. 3.11 Experimental observation of WC particle distribution in longitudinal section of an LMI layer under different conditions: **a** the Lorentz force is upward, **b** the Lorentz force is downward and **c** without the Lorentz force [9]

the longitudinal section of the LMI layer prepared from the LMI experiment under different directions of the Lorentz force induced by the electromagnetic compound field. The direction of the Lorentz force is upward in Fig. 3.11a and downward in Fig. 3.11b. The longitudinal section of the LMI layer prepared without the external Lorentz force is shown in Fig. 3.11c. The thickness of the LMI layer is almost the same for the three coating specimens, which confirms the simulation result that the electromagnetic compound field hardly influences temperature distribution for the present LMI process condition. Typical WC particle morphology in the longitudinal section of the LMI layer is illustrated in Fig. 3.12. The tungsten carbide particle remains unchanged and appears slightly decomposing. The surrounding microstructure of the particle appears hypoeutectic, which is similar to the remelting microstructure of the substrate material, as shown in Fig. 3.12a. It is implied that the carbon and tungsten content in the solidification zone are relatively low. Meanwhile, reaction layer thickness around the particle is only about 1 – 2 μm, as demonstrated in Fig. 3.12b. The above phenomenon indicates that the dissolution of WC particles is extremely limited due to the rapid solidification process during laser scanning, which is consistent with model assumptions.

To further evaluate the difference in WC particle distributions as in Fig. 3.11, all LMI layers were equally divided into three regions (upper region, middle region and bottom region). The proportions of WC particles in the corresponding regions were computed via computer graphics processing, with results attached to corresponding layers in Fig. 3.11. Under the upward Lorentz force, most particles are concentrated in the middle and the bottom region, as shown in Fig. 3.11a. Only less than 4% of WC particles are trapped in the upper region of the LMI layer, while the proportion of WC particles in the middle and bottom regions is more than 16%. In this case, the machining allowance of the LMI layer is limited because of a low concentration of

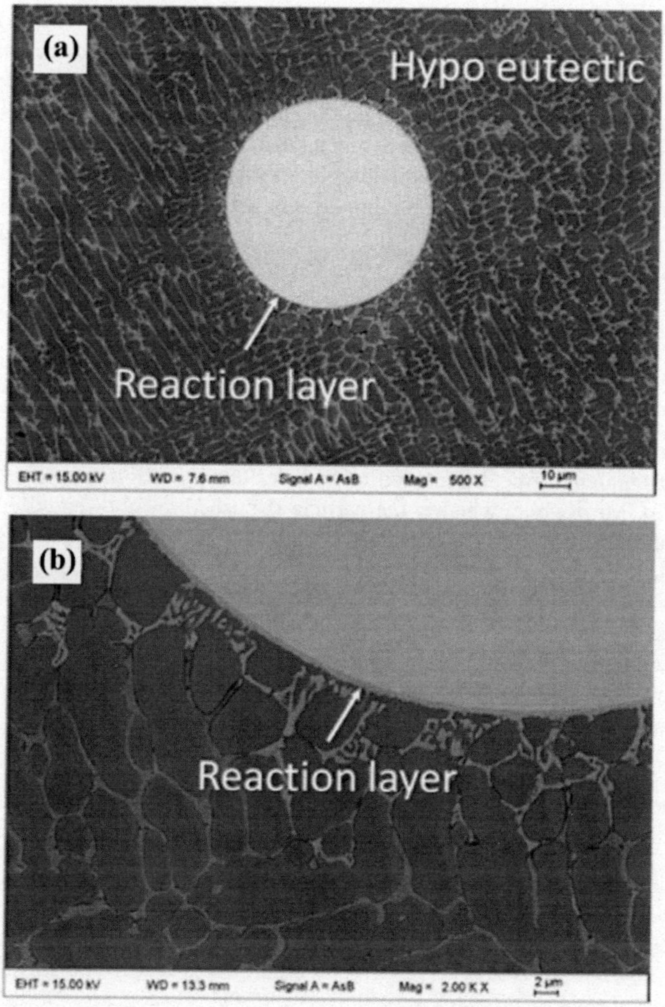

Fig. 3.12 Typical WC particle morphology in the longitudinal section of an LMI layer: **a** microstructure surrounding the particle and **b** reaction layer around the particle [9]

WC particles in the near-surface region of the LMI layer. Conversely, the majority of particles are trapped in the upper region when the direction of the Lorentz force is downward, as shown in Fig. 3.11b. The proportion of WC particles increases gradually from bottom to surface within the LMI layer. The fraction of WC particles is up to 28% in the upper region, while they are barely sunk into the lower region of the molten pool. The graded distribution of WC particles in the LMI layer exhibits several advantages, such as optimization of residual stress distribution in the coating, reduction of crack tendency, or increase of critical volume content of reinforcement particles [6]. Based on the comparison of particle distribution between the LMI condition with or without the electromagnetic compound field applied, particles are distributed relatively uniformly in the LMI layer produced without the electromagnetic compound field, as demonstrated in Fig. 3.11c. Experimental observations conform to simulation prediction, indicating that the distribution gradient of reinforcement particles can be tailored easily with the assistance of an electromagnetic compound field to match different requirements without changing the original LMI process parameters.

3.4.4 Particle Trajectory

Drag force, gravity force, buoyancy force and interfacial force all influence the pushing/engulfment transition of particles during solidification in the LMI process [23]. Meanwhile, whether the particle is pushed away or engulfed by the moving solidification interface between the liquidus and solidus phase is determined by particle size and the rate of solidification interface movement, as well as thermal conductivity of the particle. During the LMI process, the direction of particle movement near the solidification interface is downward, which is opposite to the shifting direction of the solidification interface. Therefore, particles can reach the interface and the particle/interface interaction occurs. Accordingly, equilibrium velocity V_e is given as [24]

$$V_e = \frac{d}{3\eta\lambda}\left(\frac{\Delta G_0}{R}\left(\frac{a_0}{a_0 + d}\right)^i + \frac{2R\Delta\rho g}{3\lambda}\right) \tag{3.20}$$

where d is the particle/interface separation distance, λ is thermal conductivity ratio, R is particle radius, ΔG_0 is interfacial energy difference, a_0 is atomic distance and $\Delta\rho$ is density difference.

$$\lambda = \frac{k_p}{k_m} \tag{3.21}$$

where k_p is the thermal conductivity of the particle, k_m is the thermal conductivity of the liquid. Estimated values of these parameters are $\lambda = 0.5$, $\Delta G_0 = 1$ N/m, $\Delta\rho = -7750$ kg/m^3, $g = 10$ N/kg, $K_m \approx 60$ W/(m·K), $K_p \approx 30$ W/(m·K).

$$R^* = \sqrt{3\Delta G_0\lambda/(8|\Delta\rho|g)} = 1.5 \times 10^{-3}\text{m} \tag{3.22}$$

Since 0.035×10^{-3} m $< R < 0.075 \times 10^{-3}$ m, that is, R is much smaller than R^*, Eq. (3.20) can be simplified into

$$V_e = \frac{d}{3\eta\lambda}\frac{\Delta G_0}{R}(\frac{a_0}{a_0 + d})^i \tag{3.23}$$

where $i = 2$ and $d = a_0$ [15]. Critical velocity is

$$V_c = \frac{a_0 \Delta G_0}{12\eta\lambda R} \tag{3.24}$$

hence $Vc \approx 1.0043 \times 10^{-8}/R$. The predicted variation of Vc with R is illustrated in Fig. 3.13. The diameter of WC particles used in this simulation is given as 75 – 150 μm. According to the present LMI process parameters, solidification velocity is approximately 1000 μm/s. As a result, particles can be engulfed rapidly in the solidification interface.

Differently from traditional casting, solidification velocity in the molten pool is usually very high during the laser scanning process. Once particles have reached the solidification interface, they can be engulfed and trapped immediately into it. Particle movement distance is extremely limited due to interfacial forces. Experimental observations on particle distribution also match the simulation prediction based on assumptions made in this simulation. According to the above calculation results, interfacial forces acting on particles can be neglected in this model, which is consistent with model assumptions. Meanwhile, particle distribution probability is nearly constant at the solidification interface, as indicated in Fig. 3.11c. Since it is unlikely for particles to be pushed into the position where solidification occurs at last, it can be suggested that different distributions of WC particles shown in Fig. 3.11 are caused by the additional Lorentz force.

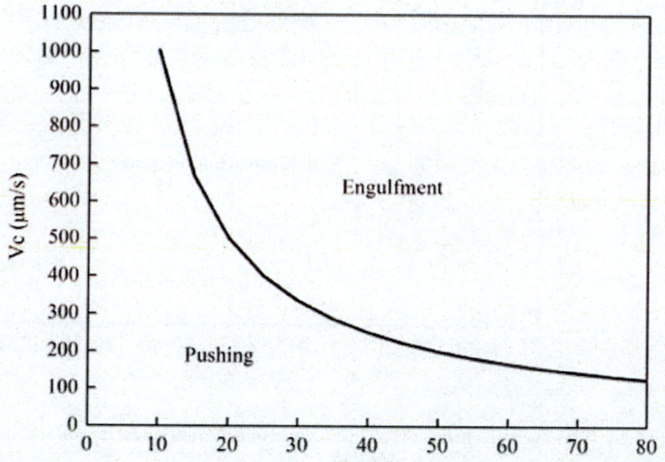

Fig. 3.13 Model-predicted variation of V_c with R [9]

Sinking trajectories are computed using the simulation model with results presented in Fig. 3.14. It is shown that these trajectories are different for particles in different injection zones of the molten pool. Particles shown in Fig. 3.14 are taken from the central zone of the laser spot. Once they have penetrated into the surface of the molten pool, they get transmitted along with the fluid motion by the strong Marangoni convection at the surface. The resulting sinking trajectories become apparently different, which is mainly attributed to the directional Lorentz force. The occurring buoyancy acting on particles increases with the downward Lorentz force. As a result, it is difficult for particles to enter the molten pool bottom zone, thus they are rapidly engulfed by the solidification interface.

Particles can penetrate the Marangoni convection quickly with the upward Lorentz force. Then they easily sink to the molten pool bottom zone, therefore they become engulfed in the bottom zone of the solidification interface. Particle motion velocity is relatively high in the case without the Lorentz force, thus the majority of particles avoid engulfment into the solidification interface due to the higher fluid velocity and drag force. With the decrease of particle velocity and the solidification interface shifting, particles will be eventually engulfed. It is demonstrated that although particle distribution can be affected by different injection positions, the overall sinking depth of particles is evidently associated with the direction of the Lorentz force.

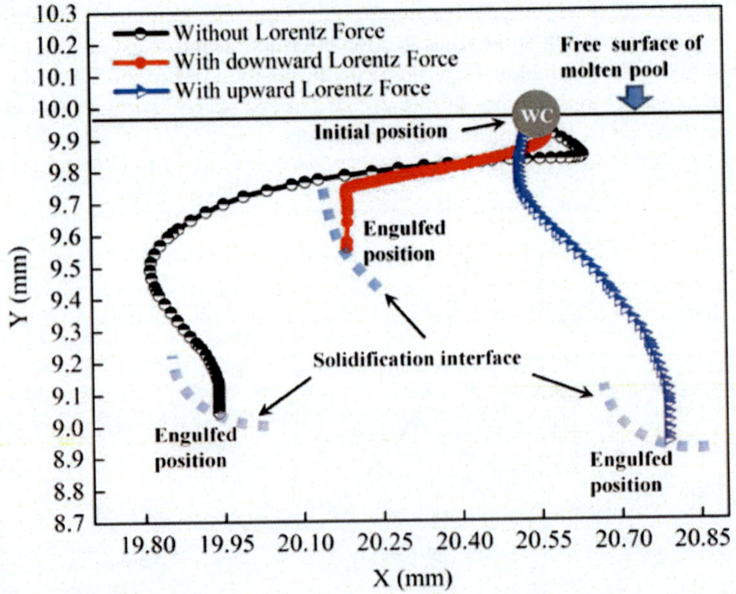

Fig. 3.14 Sinking trajectories of WC particles for different directions of the Lorentz force [9]

3.4.5 Effects of the Directional Lorentz Force

The directional Lorentz force generated by the steady magnetic field and the steady electric field is a type of directional steady volume force in the molten pool, expressed as

$$F_{Lorentz} = (\boldsymbol{j} \times \boldsymbol{B})V \tag{3.25}$$

where $F_{Lorentz}$ is the Lorentz force, and V is fluid volume displaced by a single particle. The direction of the Lorentz force acting on the fluid is in the opposite direction of the corresponding force acting on particles, as described below.

$$F_p = -F_{Lorentz} \tag{3.26}$$

where F_p is the force counteracting the Lorentz force acting on the particle. Figure 3.15 schematically shows the forces acting on a single particle and the surrounding fluid. When the Lorentz force acting on the fluid is upward, the balance equation is established to express the force (in the Y-direction) acting on the particle as follows:

$$-F_{iy} = F_{\mathrm{Buoyancy}} - F_p + F_{Darcy} - F_{Gravity} \tag{3.27}$$

where F_{iy}, $F_{Buoyancy}$, F_{Darcy} and $F_{Gravity}$ represent the inertial forces: buoyancy, drag force and gravity, respectively. Since $F_{Buoyancy}$, F_{Darcy} and $F_{Gravity}$ are assumed to be constant, F_{iy} and F_p are positively proportional. Consequently, when the direction of the Lorentz force acting on the fluid is upward, the equivalent buoyancy acting on particles is reduced, thus the sinking velocity of particles is increased. During the continuous powder injecting process, particles sink into the bottom region of the molten pool easier. Following the same principle, when the direction of the Lorentz force acting on the fluid is downward, particles will be trapped in the upper region.

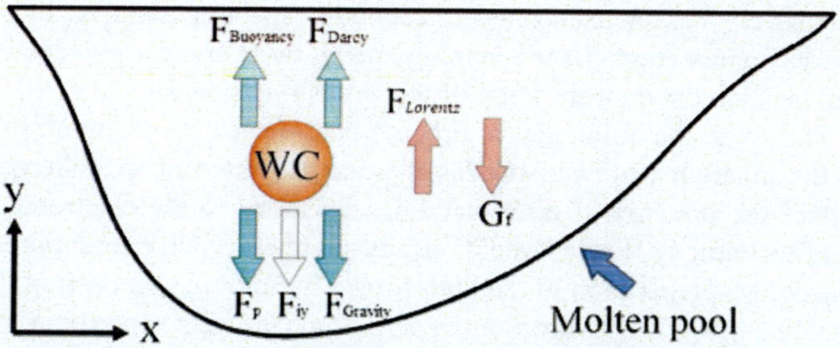

Fig. 3.15 Diagram of forces acting on the particle and the surrounding fluid [9]

Expanding the forces of $F_{Buoyancy}$ and F_p in Eq. (3.27) yields

$$(g - \frac{j \times B}{\rho})\rho V + F_{Darcy} - F_{Gravity} + F_{iy} = 0 \tag{3.28}$$

The gravity acceleration equivalent is assumed to be the resultant acceleration of local gravity, and the Lorentz force that is related to fluid density. According to Eq. (3.28), when the Lorentz force and gravity force act in the opposite direction, the gravity acceleration equivalent is decreased. The resultant buoyancy acting on particles is reduced accordingly. It can be suggested that particles in the molten pool are in hypogravity status. When the Lorentz force and gravity force act in the same direction, particles in the molten pool are in hypergravity status. In summary, the direction of the directional Lorentz force can be easily adjusted by changing the direction of the electric and magnetic field, thus controlling the movement of reinforcement particles in the LMI process.

3.5 Pore Minimization in Laser Cladding

3.5.1 Porosity in Laser Cladding

As a net-shape processing technology in the field of surface modification, additive manufacturing or remanufacturing, the laser cladding technique displays many advantages such as the formation of metallurgical bonding, low heat input, controllable scanning speed, narrow heat-affected zones and so on [25], although pores inevitably occur in the molten pool owing to the nature of the technique as well as of the powder manufacturing process. These can be triggered by various factors including trapped gas due to turbulence, release of gases from the melt, lack of fusion, etc. [26, 27]. The presence of pores deteriorates the mechanical properties of the cladding layers significantly. Impact ductility, toughness and fatigue properties may be affected in particular, thereby limiting the application of laser cladding parts in many critical and demanding conditions [28]. Therefore, research conducted extensively has focused on pore reduction in laser additive alloys by optimizing process parameters (laser scanning speed, laser energy, powder feeding rate, etc.) and the element concentration of the depositing powder [29].

However, the efficiency of process parameter optimization remains limited due to the inherent nature of the laser processing system. An alternative approach to reduce the porosity of laser cladding can resort to the electromagnetic compound field technology. For example, steady magnetic field can suppress pore initiation during the preparation of TiC reinforced aluminum composite fabricated via laser cladding [30]. Gravity drop-out of particles in the molten pool can be prevented by the application of AC magnetic field of 80 mT and 135 mT, corresponding to oscillation frequencies between 1 to 10 kHz beneath the weld specimen [31]. Furthermore, electromagnetic compound field is used to reduce the porosity of 92% AISI 316L and 8% TiH_2 composite coating [32], whereas alternating magnetic field along with DC electric field is applied to the molten pool to improve the grain structure of the 304 stainless steel cladding coating [33]. Previous studies have also focused on the application of steady, travelling and oscillatory magnetic fields aiming to obtain the required grain refinement, liquid surface, temperature distribution, acceleration or braking of the electrically-conducting fluid flow [34].

3.5.2 Laser Cladding with Electromagnetic Field

To investigate the escape behavior of pores in the molten pool during laser cladding with an external electromagnetic compound field applied, a laser cladding experiment was conducted on cast iron substrate [35]. The experimental system consists of a 2 kW high-power laser diode (Laserline, LDF400-2000), custom made electric field and steady magnetic field, as illustrated in Fig. 3.16. The steady magnetic field is produced by two electromagnetic poles placed at a distance of 20 mm with a maximum magnetic flux density of 2.0 T. The steady electric field with a high intensity DC current is applied across the molten pool to generate an electric field by a DC regulated power source (0–1000 A, 0–10 V). The distribution of magnetic flux density between the two electromagnet poles is uniform.

A nodular cast iron (QT-400) plate with dimensions of 200 × 20 × 10 mm is the substrate specimen. Figure 3.17 shows its microstructure, where *F* refers to ferrite phase and *G* refers to spherical graphite. Austenitic stainless steel powder (AISI 316L) is used as the depositing material; the physical properties of both QT-400 and AISI 316L are given in Table 3.4 [36]. Laser power density, scanning speed and powder-feeding rate in the experiment are 3.58×10^7 W/m^2, 16 mm/s and 20 g/min, respectively. The focal length measures approximately 270 mm, with the focal spot located on the surface of the substrate; powder size varies between 50 – 125 μm. During the deposition process, powder is injected into the laser cladding region by means of a coaxial powder-feeding system. High-purity argon is used as a shielding gas at a flow rate of 10 L/min, and high-purity nitrogen gas as a powder feeding gas at a working rate of 9.4 L/min. Figure 3.18 schematically displays the geometrical features of a single clad layer. The cross-section morphology of this layer was examined through an optical microscope (OM) (CarlZeiss, Axio Scope A1). Defects of the cladding layer were investigated by OM (CarlZeiss, Axio Scope A1), whereas microstructure was analyzed by SEM (Carl Zeiss, EVO 18).

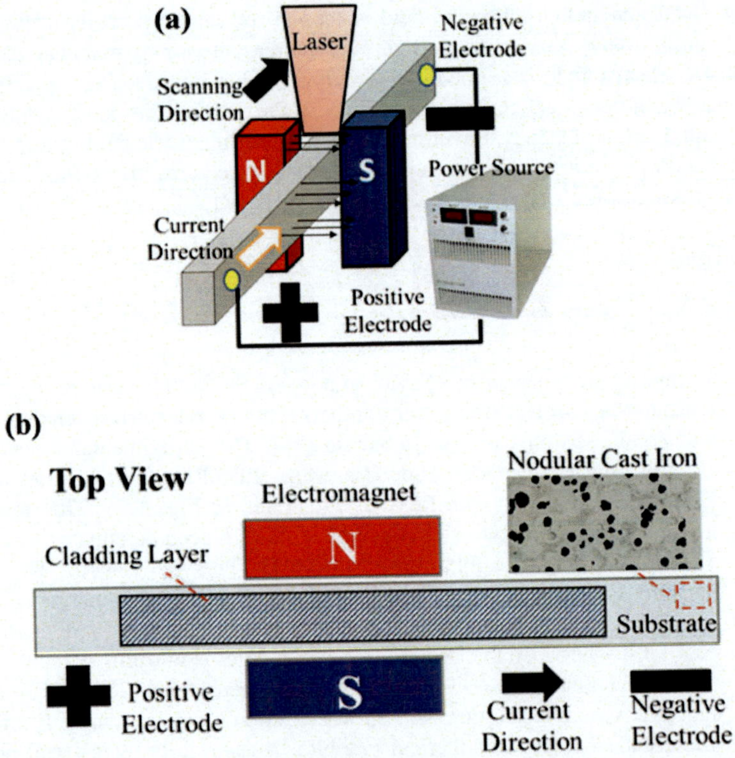

Fig. 3.16 Diagram of the laser cladding experiment setup: **a** 3D review and **b** top view [35]

Fig. 3.17 Nodular cast iron (QT-400) microstructure [35]

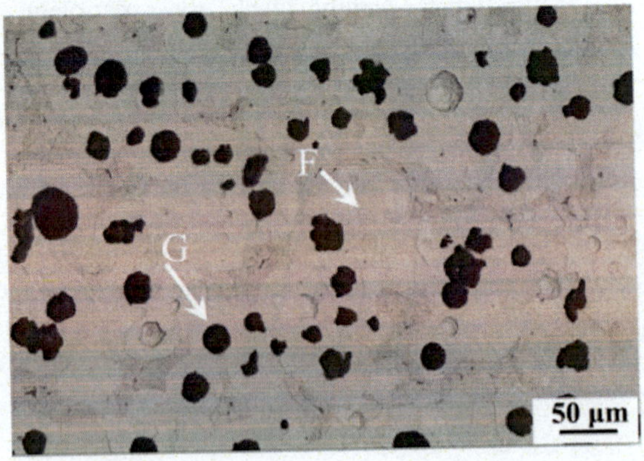

Table 3.4 Physical properties of the substrate and the coating materials [36]

Properties	Substrate cast iron	Powder AISI 316L
Solid density ρ_s (kg m^{-3})	7090	7800
Liquid density ρ_l (kg m^{-3})	6600	7800
Solidus temperature T_s (K)	1393	1650
Liquidus temperature T_l (K)	1498	1700
Liquid specific heat cp_l (J kg^{-1} K^{-1})	704	746
Latent heat of fusion (J kg^{-1})	8.7	30
Dynamic viscosity η (Pa s)	5.5×10^{-3}	0.006
Temperature coefficient of surface tension (N m^{-1} K^{-1})	-4.3×10^{-4}	-5.2×10^{-4}
Thermal expansion coefficient (1/K)	1.35×10^{-5}	1.30×10^{-5}

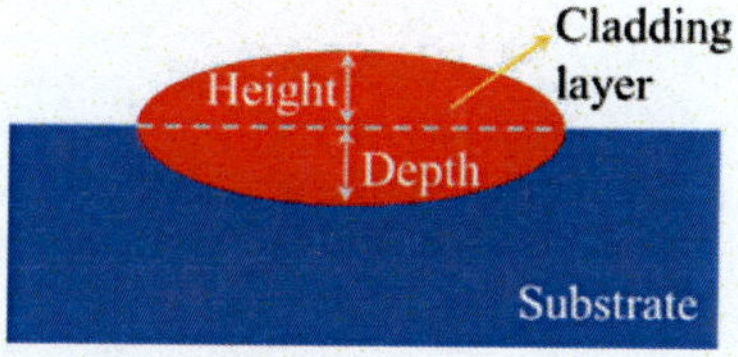

Fig. 3.18 Schematic geometry scheme of a single clad layer [35]

3.6 Simulation of Electromagnetic Field-Assisted Laser Cladding

3.6.1 Assumptions and Molten Pool Modeling

With regards to the laser cladding experiment, the commercial software COMSOL Multiphysics 5.3® was used to model transient heat transfer, fluid flow, electromagnetic compound field, and pore movement in the molten pool. A 2D molten pool model with laser and an electro-magnetic field applied was created with 40 mm length and 10 mm height. The following assumptions were made within the simulation framework to simplify the numerical computation:

(1) The liquid metal is considered an incompressible Newtonian fluid, and the flow within the molten pool is laminar [37].
(2) The solid–liquid phase transformation is modeled by the enthalpy-porosity approach, and the liquid fraction is treated as the function of temperature [38].
(3) The initial temperature of the work piece and surrounding environment is set to 293 K.
(4) The surrounding gas phase is not modeled due to the large difference in density and dynamic viscosity between the liquid metal and the gas phase [22].

(5) The effect of temperature rise due to Joule heating is ignored, as the heating time of current is very short (5 s) and temperature rise is only approximately 19 °C by the influence of Joule heating.

The model is formulated based on mass conservation Eq. (3.1) and momentum conservation as

$$\rho\frac{\partial \boldsymbol{u}}{\partial t} + \rho(\boldsymbol{\mu}\cdot\nabla)\boldsymbol{u} = -\nabla p + \nabla\cdot[\eta(\nabla\boldsymbol{u} + (\nabla\boldsymbol{u})^T) - \frac{2}{3}\eta(\nabla\cdot\boldsymbol{u})I] + \boldsymbol{F}_{Buoyancy} + \boldsymbol{F}_{Darcy} + \boldsymbol{F}_{Lorentz} \tag{3.29}$$

$$\boldsymbol{F}_{Buoyant} = \rho g\beta(T - T_m) \tag{3.30}$$

$$\boldsymbol{F}_{Darcy} = C_1\frac{(1-f_L)^2}{f_L^3 + C_2}\boldsymbol{u} \tag{3.31}$$

where C_1 and C_2 are coefficients of the Darcy source term. In the molten pool, the Lorentz force is treated as a type of volume force to couple with the liquid metal and is obtained by the cross multiplication of electric current density and magnetic flux density, expressed as

$$F_{Lorentz} = \boldsymbol{j}\times\boldsymbol{B} = \varepsilon(\boldsymbol{E} + \boldsymbol{u}\times\boldsymbol{B})\times\boldsymbol{B} = \boldsymbol{j}_1\times\boldsymbol{B} + \varepsilon(\boldsymbol{u}\times\boldsymbol{B})\times\boldsymbol{B} \tag{3.32}$$

where$\boldsymbol{j}_1$ is the electric flux density applied; the term $\boldsymbol{j}_1\times\boldsymbol{B}$ is the directional Lorentz force generated by the external current and magnetic field; the term $\varepsilon(\boldsymbol{u}\times\boldsymbol{B})\times\boldsymbol{B}$ is the induced Lorentz force generated by the movement of electrically conducting media perpendicular to the magnetic field. Current conservation is expressed as.

$$\nabla\cdot\boldsymbol{j}_1 = 0 \tag{3.33}$$

Energy conservation is given as.

$$\rho C_p\frac{\partial T}{\partial t} + \rho C_p\boldsymbol{u}\cdot\nabla T = \nabla\cdot(k\nabla T) \tag{3.34}$$

3.6.2 Modeling of Molten Pool with Pores

The process of diffusion of solute elements in the molten pool is not considered while modeling pore movement during laser cladding; growth, geometric deformation and mutual fusion of pores are not taken into account, either. Surface tension force can hinder pores from detaching during their growth [39]. The departure diameter of pores at the solid–liquid interface was found to be approximately 100 – 1000 μm [40]; small pores escape less easily. In this simulation, a pore diameter of 100 μm is assumed at the beginning. Pore mass is calculated based on carbon monoxide (CO) density. Pore motion is simulated using the particle tracing module of COMSOL Multiphysics 5.3® software. In this software, a pore is assumed to be a virtual particle. When the molten pool fluid is subjected to the directional Lorentz force, pore motion is defined by Newton's second law combined with the electromagnetic compound field according to the action-reaction relationship between fluid and pore, as

$$\frac{d(m_p U_p)}{dt} = F_{Drag} + F_{Gravity} + F_{Buoyant} - F_{Lorentz} \cdot V_{\text{pore}} \tag{3.35}$$

where m_p is pore mass, U_p is pore velocity, F_{Drag} is drag force, F_{Gravity} is pore gravity, F_{Buoyancy} is pore buoyancy, F_{Lorentz} is the force produced by the directional Lorentz force and V_{pore} is pore volume.

The first step of this simulation is to compute the flow field; the trajectory and velocity of pores can then be calculated by substituting data on position, velocity and electromagnetic field at each node of the molten pool model into the motion equation of the pore. Forces acting on the pore are schematically shown in Fig. 3.19 including buoyancy, gravity, the Lorentz force and the surrounding fluid resistance (drag force) acting on the pores. According to the expression for the steady drag of an unbounded uniform flow over a spherical bubble with $0 \leq Re \leq 1000$, obtained by Mei and Klausnethe [40], the drag force is formulated as

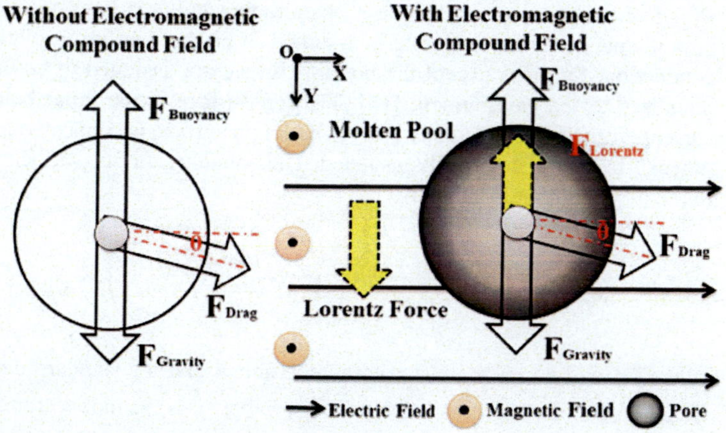

Fig. 3.19 Diagram of force analyses on the pore [35]

$$\frac{F_{Drag}}{3\pi\rho\eta u d_b} = \frac{2}{3} + \left[\left(\frac{12}{Re}\right)^{0.65} + 0.796^{0.65}\right]^{-1/0.65} \quad (3.36)$$

$$Re = \frac{u d_b}{\eta} \quad (3.37)$$

where d_b is pore diameter and Re is the Reynolds number.

3.6.3 Boundary Conditions and Model Parameters

Figure 3.20 describes boundary conditions for the numerical model. The surface tension gradient is considered to be the original driving force of molten pool flow. The boundary condition is governed by shear stress in the liquid at the molten pool surface, which is directly related to the surface tension gradient [36]. The expression for this boundary condition is given by Eqs. (3.15) and (3.16). The energy distribution intensity I_0 (W·m^{-2}) of laser is considered to be uniform, and laser source movement is added as the boundary heat flux at the top of the solution domain. It is defined as

$$I_0(x,t) = \begin{cases} \frac{\alpha p_l}{\pi r_l^2} & |x - v_s t| \le r \\ 0 & |x - v_s t| > r \end{cases} \quad (3.38)$$

where α is the laser absorptivity of AISI 316L to diode laser (wavelength = 1.06 μm), P_l is incident laser power, $r_l$is laser beam radius and v_s is laser beam velocity. The bottom surface of the substrate is assumed to be the thermal insulation boundary conduction. Surfaces except the bottom surface are assumed to be the convective boundary to the environment [15]. The left surface is the input boundary of the current with a current value of 500 A. The right surface is connected to earth, while other surfaces are electrically insulated. The boundary moving velocity V_P [22] due to the powder input is given by

$$V_p = N_p \frac{\eta_w D_m}{\rho_w \pi r_p^2} exp\left(-N_p \frac{(x - V_s \cdot t)^2}{r_p^2}\right) \times \vec{j} \quad (3.39)$$

where N_p and r_p are the constriction coefficient and the standard deviation of the Gaussian distribution, respectively. The variable $\vec{j}$ is the unit vector in Y-direction, D_m is powder feeding rate. Powder catchment efficiency is expressed as η_w. Powder density is ρ_w at $T = T_m$. The Deformation Geometry method is used to track the motion of the free interface in COMSOL Multiphysics 5.3® software using the Winslow smoothing approach [21]. The boundary condition for controlling the free surface is defined by Eq. (3.19).

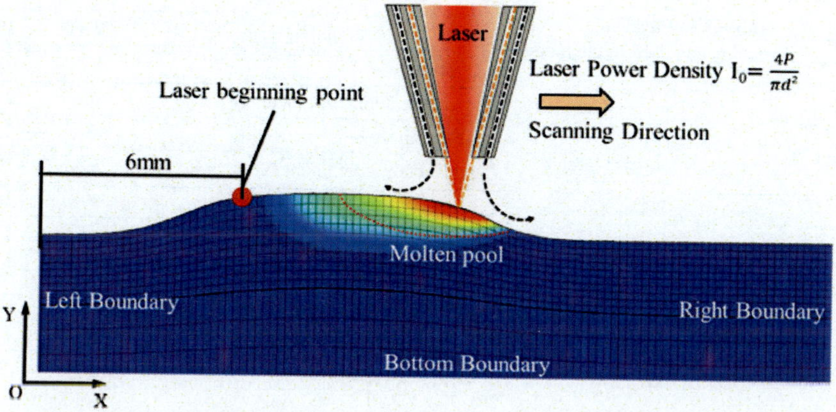

Position	Top	Left	Right	Bottom
Heat transfer	$\lambda \cdot \vec{\nabla} T \cdot \vec{n} = -h_c(T - T_0) - \varepsilon\sigma(T^4 - T_0^4)$	$\lambda \cdot \vec{\nabla} T \cdot \vec{n} = -h_c(T - T_0) - \varepsilon\sigma(T^4 - T_0^4)$	$\lambda \cdot \vec{\nabla} T \cdot \vec{n} = -h_c(T - T_0) - \varepsilon\sigma(T^4 - T_0^4)$	$-\vec{n} \cdot q = 0$
Fluid	Normal direction: $F_\sigma \cdot \vec{n} = -\kappa\gamma \cdot \vec{n}$ Tangential direction: $F_\sigma \cdot \vec{t} = \frac{\partial \gamma}{\partial T} \vec{\nabla} T \cdot \vec{t}$	$\vec{u} = 0$	$\vec{u} = 0$	$\vec{u} = 0$
Moving mesh	$V_n = \vec{u} \cdot \vec{n} + \vec{V}_p \cdot \vec{n}$	$V_n = 0$	$V_n = 0$	$V_n = 0$
Current	$\vec{n} \cdot j = 0$	$\vec{n} \cdot j = j_0$	$V_{Voltag} = 0$	$\vec{n} \cdot j = 0$

Fig. 3.20 Summary of boundary conditions for the numerical model [35]

The governing equations of the model with the mesh consisting of 5000 quadrilateral elements were solved using the COMSOL Multiphysics 5.3® software. By reducing the mesh size through iterating process to a maximum mesh size of 0.3 mm, simulation results were not affected. The direct PARDISO solver associated with generalized-α temporal solver was used and the time step was automatically set by the solver. The physical parameters used in this model are given in Table 3.5; addi-

Table 3.5 Physical parameters used in the model [35]

Initial temperature (K)	T_0	293
Emissivity	ε	0.4
Powder catchment efficiency	η_p	0.45
Absorptivity	α	0.4
Convection coefficient (W m^{-2} K^{-1})	h_c	20
Laser power (W)	P_l	1800
Laser beam radius (mm)	r_l	4
Mass powder rate (g min^{-1})	D_m	20
Powder stream radius (mm)	r_p	2.2
Constriction coefficient	N_p	5
Scanning speed (mm s^{-1})	V_s	12

tional physical properties of the substrate material (nodular cast iron) are illustrated in Fig. 3.21 as a function of temperature.

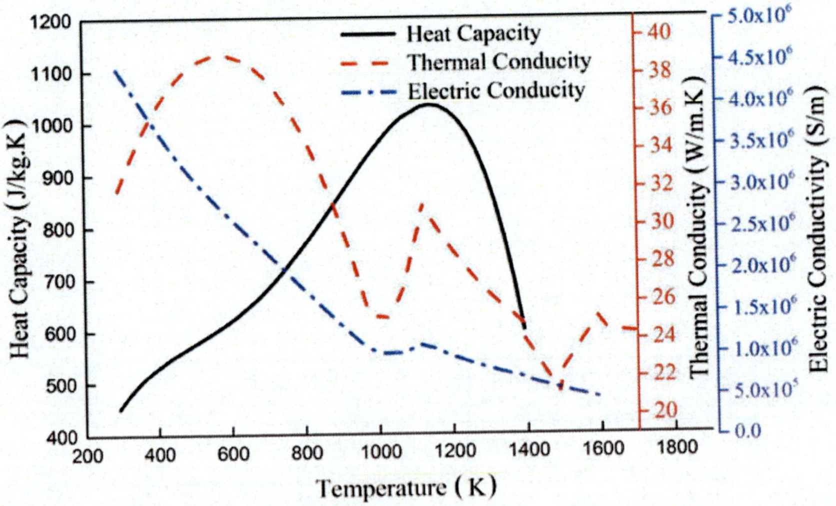

Fig. 3.21 Physical properties of nodular cast iron [35]

3.7 Molten Pool Behavior

3.7.1 Longitudinal Section of Cladding Layers

Longitudinal sections of the cladding layers processed under different conditions of external electromagnetic field along with simulation predictions are shown in Fig. 3.22. The dotted line corresponds to molten pool shape obtained through the simulation. Four sets of electromagnetic field parameters are used in the model: (1) without magnetic field, (2) with steady magnetic field alone, (3) with the downward Lorentz force and (4) with the upward Lorentz force. The model uses the same process parameters (scanning speed of 16 mm/s, powder feeding rate of 20 g/min, laser power density of 3.58×10^7 W/m^2) as the laser cladding experiment. The direction of the Lorentz force is represented by *MN*. When $MN = 1$, the direction of the Lorentz force is upward, which is opposite to the direction of gravity. When $MN = -1$, the direction of the Lorentz force is downward, which is the same as the direction of gravity. It is clear that the surface gas–liquid interface curve and the solid–liquid interface curve of the molten pool comply with experimental results

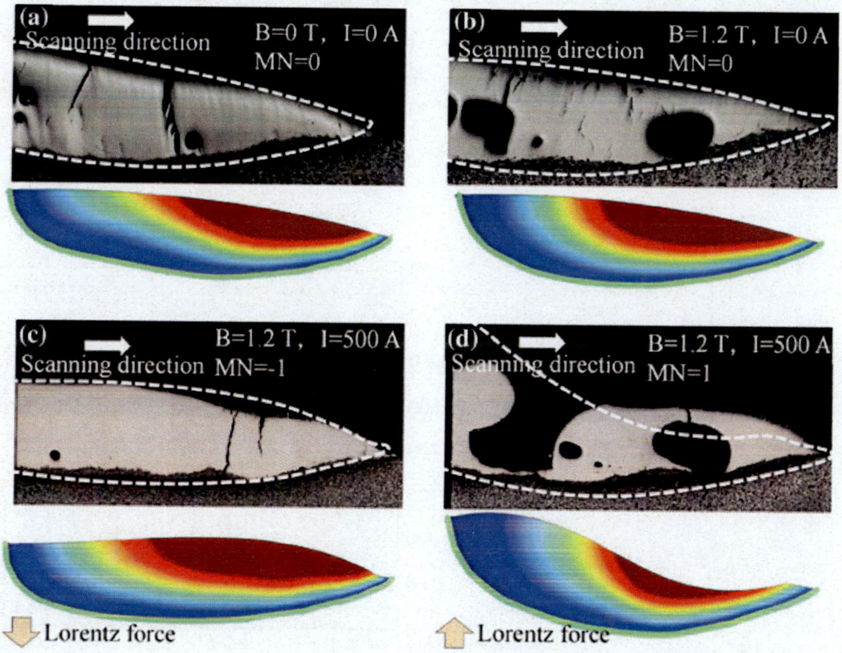

Fig. 3.22 Longitudinal sections of cladding layers obtained under different conditions of external electromagnetic compound field: **a** $B = 0$ T, $I = 0$ A, $MN = 0$; **b** $B = 1.2$ T, $I = 0$ A, $MN = 0$; **c** $B = 1.2$ T, $I = 500$ A, $MN = -1$; **d** $B = 1.2$ T, $I = 500$, $MN = 1$ [35]

for different electromagnetic conditions, indicating that simulations of the cladding process and the temperature field are highly accurate.

For a better validation of the model, the height and depth of the cladding layer in longitudinal section are compared between the simulation and experimental results, as shown in Fig. 3.23. Both reveal the same changing trend in cladding layer height and depth, such that (1) a steady magnetic field has little effect on molten pool height and depth; (2) the upward Lorentz force contributes to the increase in cladding layer height and vice versa; (3) the electromagnetic compound field has little effect on cladding layer depth. The error in cladding layer geometrical dimensions between experimental and simulation results is less than 10%. Therefore, model predictions are in good agreement with experimental results.

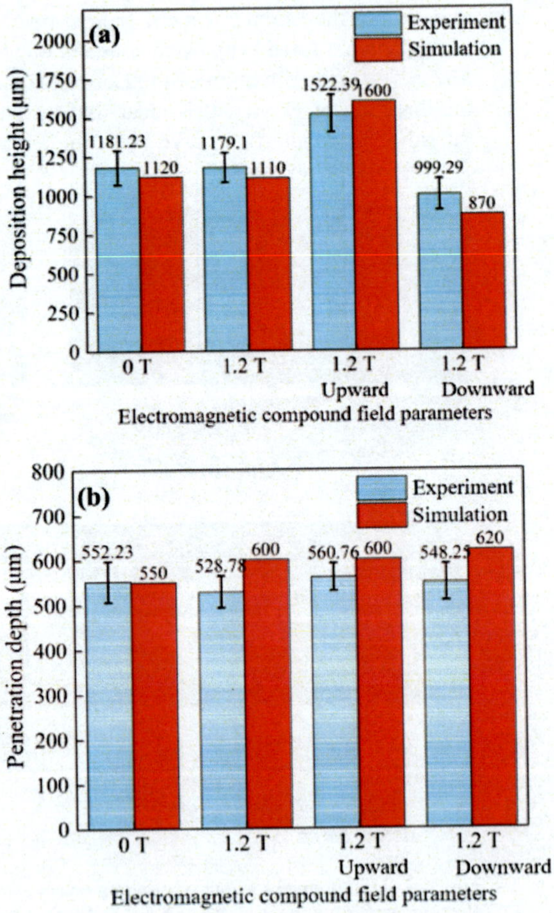

Fig. 3.23 Comparison of **a** deposition height and **b** penetration depth between simulation and experimental results [35]

3.7.2 Lorentz Force Performance

Distributions of the Lorentz force in the longitudinal section of the molten pool under different intensities of electromagnetic compound field are shown in Fig. 3.24, where the green line indicates the molten pool boundary, the red arrow denotes the direction of the Lorentz force and the blue arrow indicates fluid flow direction. As demonstrated in Fig. 3.24a, the peak Lorentz force occurs at the molten pool surface

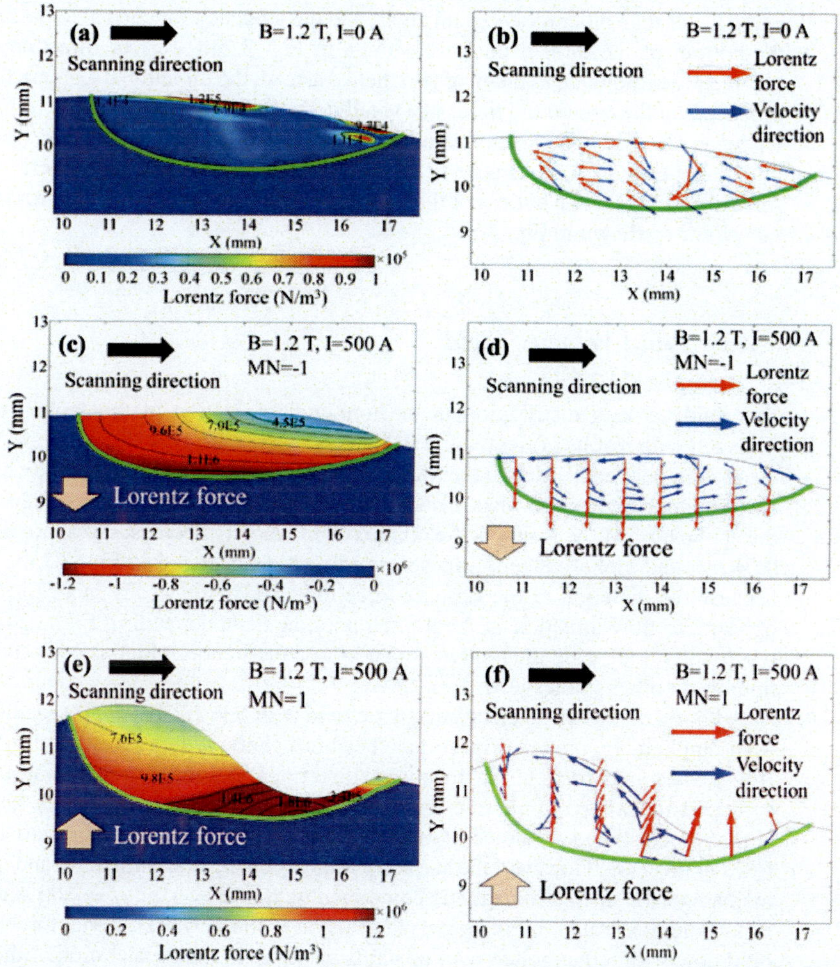

Fig. 3.24 Distribution of the Lorentz force under different conditions of electromagnetic compound field: (**a**), **b** $B = 1.2$ T, $I = 0$ A; (**c**), **d** $B = 1.2$ T, $I = 500$ A, $MN = -1$; **e**, **f** $B = 1.2$ T, $I = 500$ A, $MN = 1$ [35]

with an average magnitude of approximately 10^4 N/m^3 under a magnetic field of 1.2 T. The direction of this force is opposite to fluid flow in the molten pool, as shown in Fig. 3.24b. When the electromagnetic compound field parameters are B = 1.2 T, I = 500 A and MN = −1, two types of the Lorentz force are produced simultaneously, that is, the Lorentz force induced as shown in Fig. 3.24a, and the directional Lorentz force as displayed in Fig. 3.24c. Compared with the effect of a steady magnetic field, the directional Lorentz force is about 10 to 100 times larger than the induced Lorentz force. Moreover, the maximum of the Lorentz force occurs at the bottom of the molten pool, which is due to temperature near the bottom interface being lower than other molten pool areas, and the tendency for current to pass across the bottom of the molten pool. As shown in Fig. 3.24d, current direction is not completely orthogonal to the magnetic field. Instead, the directional Lorentz force is produced by the compound fields at a small deflective angle to the gravity direction. Under an electromagnetic compound field of B = 1.2 T, I = 500 A and MN = 1 the directional Lorentz force is upward, as shown in Fig. 3.24f, and the highest Lorentz force also occurs at the bottom of the molten pool in the magnitude of approximately 10^6 N/m^3, as shown in Fig. 3.24e.

3.7.3 *Fluid Velocity Field*

The fluid velocity distribution in the longitudinal section of the molten pool is analyzed with results presented in Fig. 3.25. Surface fluid flows outward away from the molten pool center due to the Marangoni convection, with fluid velocity near the surface much higher than that inside the molten pool, as shown in Fig. 3.25a. With the application of the magnetic field, maximum velocity decreases to approximately 0.04 m/s and global velocity is also suppressed, as depicted in Fig. 3.25b. This is because the direction of the Lorentz force induced by the steady magnetic field is opposite to flow direction in the molten pool, as demonstrated in Fig. 3.24b. As a result, the force inside the molten pool acts as a resistance, which is consistent with findings in other studies [15].

Additionally, when an electromagnetic field with B = 1.2 T, I = 500 A and MN = −1 is applied, the molten pool surface becomes gentle with a severe vortex formed (Fig. 3.25c). In particular, the downward directional Lorentz force promotes natural convection as a type of volume force such as gravity. Moreover, surface geometry is mainly attributed to the combined effect of gravity and the downward Lorentz force in this case. However, the surface geometry changes significantly and velocity increases with an electromagnetic compound field of B = 1.2 T, I = 500 A and MN = 1, as demonstrated in Fig. 3.25d. This is due to the upward Lorentz force driving fluid flow within the molten pool upwards resulting in the acceleration of the liquid metal.

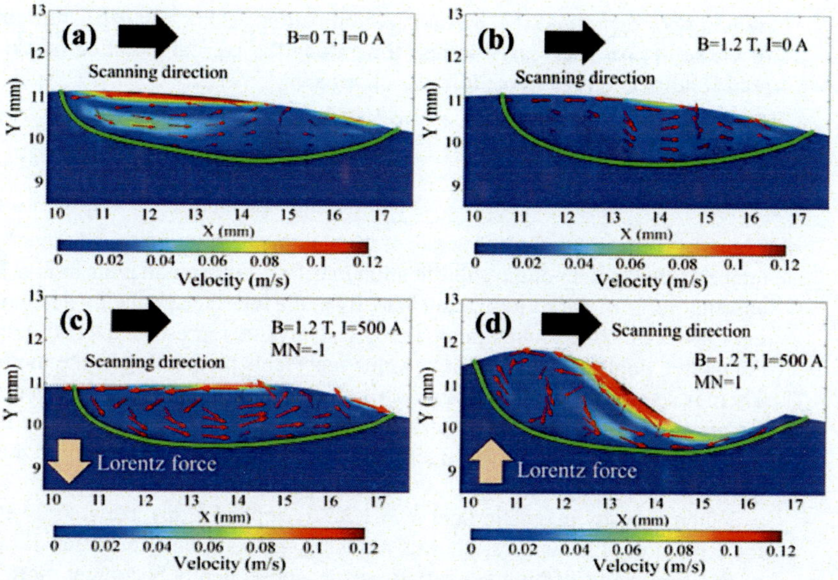

Fig. 3.25 Fluid velocity distributions under different conditions of electromagnetic compound field: **a** $B = 0$ T, $I = 0$ A; **b** $B = 1.2$ T, $I = 0$ A; **c** $B = 1.2$ T, $I = 500$ A, $MN = -1$; **d** $B = 1.2$ T, $I = 500$, $MN = 1$ [35]

3.7.4 Solidification Velocity

Figure 3.26 displays the solid–liquid interface solidification velocity in the normal direction under different electromagnetic field conditions. The Y-axis represents

Fig. 3.26 Solidification velocity profiles for normal direction under different conditions of electromagnetic compound field conditions [35]

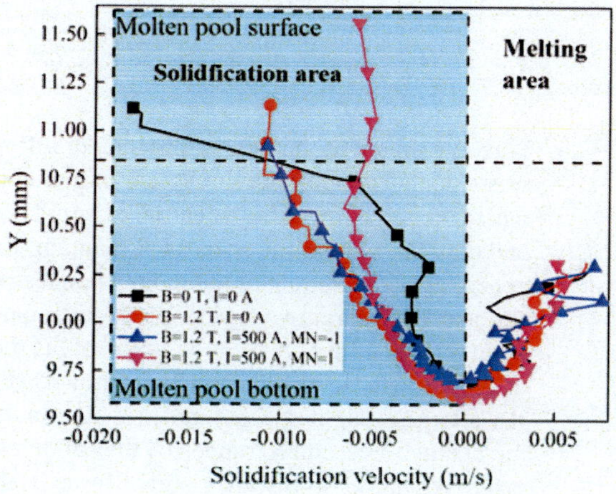

molten pool depth and the x-axis is solidification velocity, which is the thickness of the solid phase increased in unit time along the normal direction of the molten pool solidification interface. It is calculated as

$$\frac{dx_n}{dt} = \frac{dT_n}{dt} \cdot \frac{dx_n}{dT} \tag{3.40}$$

where $\frac{dT_n}{dt}$ is the normal cooling rate of the solidification interface, whereas $\frac{dx_n}{dt}$ is the temperature gradient of the solidification interface. As seen on plots in Fig. 3.26, the absolute value of the solidification velocity at the interface gradually decreases from top to bottom in the molten pool. The negative value represents the solidification of the molten pool, while the positive value represents the melting of the molten pool. There is a sudden change in solidification velocity at the surface of the molten pool for the condition without an external field. This is mainly due to the fact that fluid inside the molten pool moves vigorously, and surface heat is drawn into the molten pool.

When a steady magnetic field $B = 1.2$ T is applied alone, the global velocity of the molten pool is suppressed, thus reducing the convective heat transfer capacity of the fluid. Solidification velocity increases subsequently becoming more uniform along the solid–liquid interface. When the downward Lorentz force is applied, it gets very close to the velocity of the steady magnetic field. When the Lorentz force is upward, heat cannot travel easily within the molten pool by convection, since the surface fluid moves upward under the action of upward bulk force resulting in the interface solidification velocity severely suppressed. From simulation results it can be concluded that the electromagnetic compound field increases solidification velocity, which may in turn influence the elimination of pores from the molten pool.

3.8 Pore Escape from the Molten Pool

3.8.1 Effects of the Lorentz Force

Pore movement in the molten pool under different magnitudes of the Lorentz force was calculated using the numerical model with results are presented in Fig. 3.27, where the red arrow is the direction of liquid metal movement, the red dot is the pore and the black line is the pore track diagram. In the process of laser cladding, pores generated at the bottom of the molten pool escape less easily than others close to the surface. Therefore, two typical positions point A (X = 16 mm, Y = 10.5 mm) and point B (X = 17.5 mm, Y = 10.5 mm) at the molten pool bottom are chosen for data collection at the moment of 1.37 s, as shown in Fig. 3.27a. The pore trajectory does not rise along a straight line, but moves in an annular shape under the action of the liquid metal drag force, and is trapped by the solidification interface to settle in the cladding layer. When the downward Lorentz force is applied, as shown in Fig. 3.27b, pores

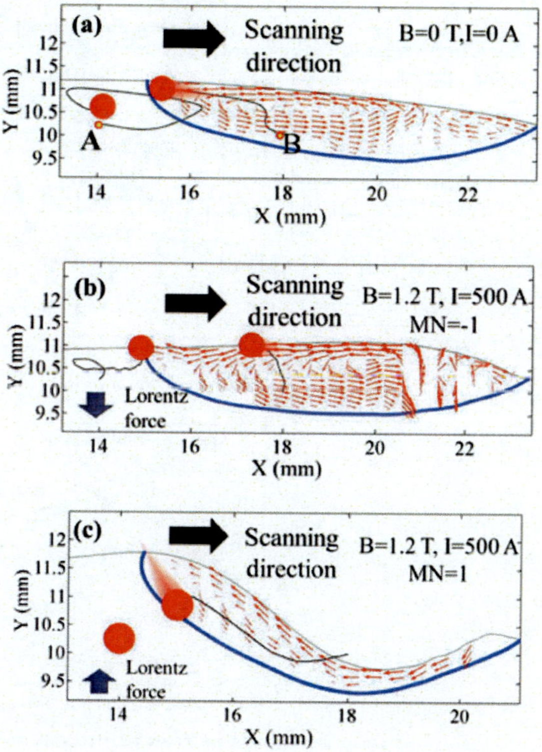

Fig. 3.27 Pore movements under different electromagnetic compound field conditions: **a** $B = 0$ T, $I = 0$ A; **b** $B = 1.2$ T, $I = 500$ A, $MN = -1$; **c** $B = 1.2$ T, $I = 500$ A, $MN = 1$ [35]

both at point *A* and *B* can escape from the molten pool. When the upward Lorentz force is applied, the pore at point *A* is captured by the solidification interface and no motion occurs. The pore at point *B* is initially dragged by the upward movement of metal fluid and is finally captured by the solidification interface, as displayed in Fig. 3.27c.

Figure 3.28 displays pore movement trajectory diagrams and the Y-direction velocity of pores under different electromagnetic field conditions. When the downward Lorentz force is applied, pores are free of the drag effect from the surface Marangoni flow, and can escape successfully with a magnetic flux density of 1.2 T. The downward Lorentz force enhances the ability of pores to move upwards, and their motion becomes more upward with the increase of magnetic flux density, as illustrated in Fig. 3.28b. When the Lorentz force turns to upward, an extreme situation occurs such that the direction of pore movement reverses back towards the molten pool bottom.

In order to further investigate the effect of the Lorentz force on the control of pore movement velocity, the duration of pore escape from the liquid under different electromagnetic field conditions was determined, as reported in Table 3.6. At point *A*, only when the downward Lorentz force is applied with a magnetic flux density of

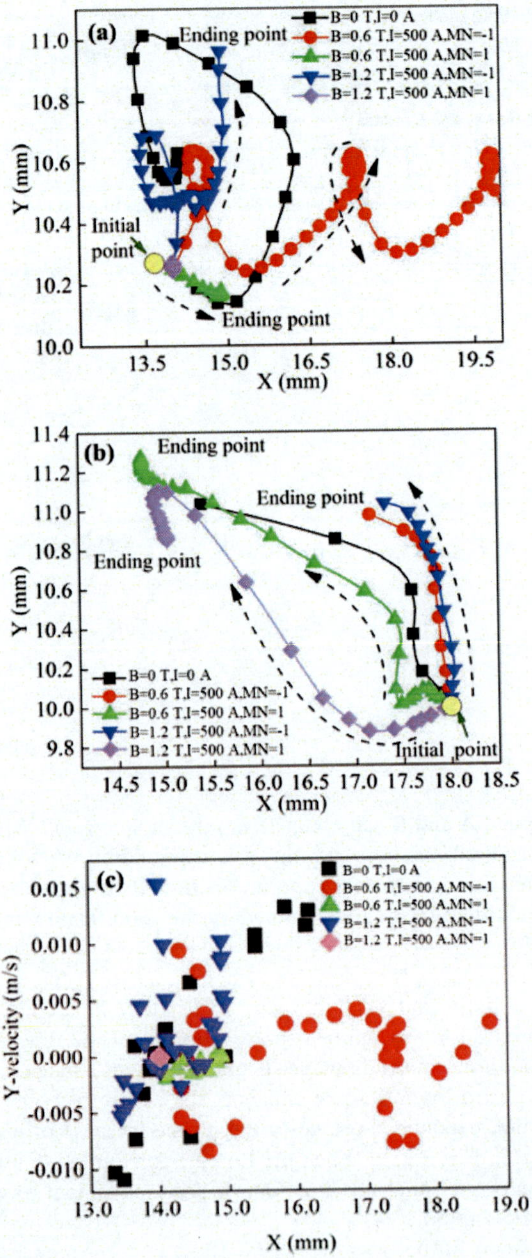

Fig. 3.28 Pore movement profiles for different electromagnetic field conditions in the position of **a** point *A*, and **b** point *B*; **c** Y-direction velocity of the pore at point *A*; **d** Y-direction velocity of the pore at point *B* [35]

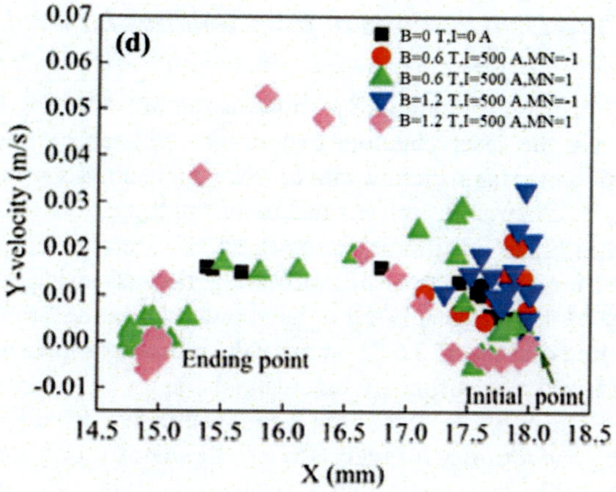

Fig. 3.28 (continued)

Table 3.6 Duration of pore escape from the molten liquid under different electromagnetic field conditions [35]

Condition	Duration	
	Point A	Point B
B = 0 T, I = 0 T	Inside the molten pool	0.119 s
B = 0.6 T, I = 500 A, MN = −1	Inside the molten pool	0.079 s
B = 0.6 T, I = 500 A, MN = 1	Inside the molten pool	0.219 s
B = 1.2 T, I = 500 A, MN = −1	0.339 s	0.079 s
B = 1.2 T, I = 500 A, MN = 1	Inside the molten pool	Inside the molten pool

1.2 T, the pore can escape from the molten pool successfully, with the length of escape of approximately 0.339 s. At point *B*, when the downward Lorentz force is applied, the length of escape of the pore decreases from 0.119 s to 0.069 s with the magnetic flux density increasing from 0 to 1.2 T, indicating that the downward Lorentz force has an ability to accelerate pore escape. When the Lorentz force direction is changed to upward with a magnetic flux density of 0.6 T, the pore escapes at 0.219 s, which is longer than that under the other two conditions (with the downward Lorentz force and without an external field). When the magnetic field is increased to 1.2 T, the pore is kept inside the molten pool, therefore, the upward Lorentz force has an effect of preventing the pore from floating.

3.8.2 Pore Distributions in the Cladding Layer

Figure 3.29 displays longitudinal microstructures of cladding layer specimens produced via the laser cladding experiment with and without an external electromagnetic compound field applied. Pore concentrations for different intensities of magnetic, electric and electromagnetic fields were calculated with the results illustrated in Fig. 3.30. In the case without an external electromagnetic compound field applied, a large number of pores are generated inside the cladding layer with a porosity of approximately 20%, as shown in Fig. 3.29a. Under the low steady magnetic flux density of 0.6 T, porosity decreases to approximately 16% with pore distribution and size still nearly unchanged (Fig. 3.29b). However, the number and diameter of the pores decrease with the magnetic flux density increased to 1.2 T (see Fig. 3.29c), and porosity is reduced to 6%. In order to compare this with the effect of the downward Lorentz force, and to further investigate the influence of the Lorentz force direction on cladding layer porosity, the upward Lorentz force is coupled to the molten pool. When the upward Lorentz force with a magnetic flux density of 0.6 T and current of 500 A is applied, two types of pores: spherical-like pores and larger open pores can be found in the cladding layer (Fig. 3.29d). Large-size open pores are generally observed (Fig. 3.29e) in the cladding layer with the magnetic field intensity increased to 1.2 T, whereas the number and size of pores are both depressed under the downward Lorentz force, as observed in Figs. 3.29f, g. In particular, a nearly perfectly dense cladding layer can be achieved under a high magnetic flux density of 1.2 T (Fig. 3.29g). Since the electromagnetic compound field is composed of a magnetic and an electric component, electric fields in two opposite directions are applied to exclude the effect of current. The results show that pores still remain in the cladding layer with the porosity staying almost the same, as shown in Fig. 3.29h, i. Consequently, it is proved that the effect of current on pore escape is negligible. The influence of an external electromagnetic compound field on cladding layer porosity is summarized in Fig. 3.30.

3.8.3 Pore Formation Mechanism

Gas pores can form in the molten pool in the process of laser cladding nodular cast iron by the following four main mechanisms. Firstly, oxygen in the molten pool reacts with carbon to produce CO gas:

$$[C] + [O] = CO \tag{3.41}$$

Secondly, when surface metal is oxidized, the resulting oxide inclusions can be transported into the molten pool. The metal oxide can then react with the carbon dissolved in the molten pool. Consequently, more CO gas is released into the molten pool according to the following reactions:

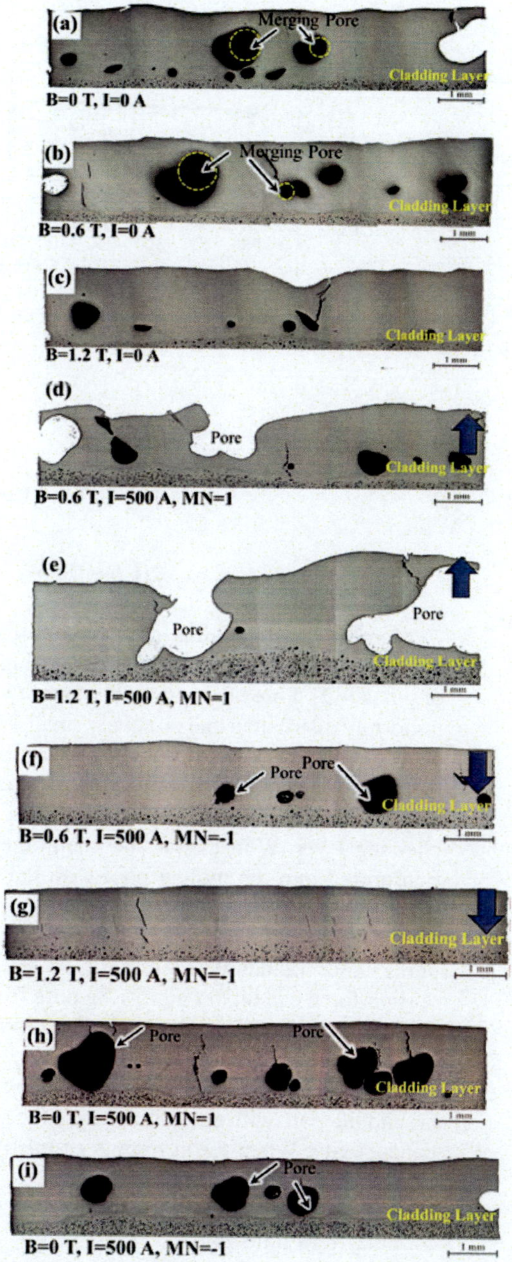

Fig. 3.29 Pore distributions in longitudinal section of cladding layer fabricated under external electromagnetic compound field applied with different parameters: **a** $B = 0$ T, $I = 0$ A, $MN = 0$; **b** $B = 0.6$ T, $I = 0$ A, $MN = 0$; **c** $B = 1.2$ T, $I = 0$ A, $MN = 0$; **ed** $B = 0.6$ T, $I = 500$ A, $MN = 1$; $B = 1.2$ T, $I = 500$ A, $MN = 1$; **f** $B = 0.6$ T, $I = 500$ A, $MN = -1$; **g** $B = 1.2$ T, $I = 500$ A, $MN = -1$; **h** $B = 0$ T, $I = 500$ A, $MN = 0$; **i** $B = 0$ T, $I = -500$ A, $MN = 0$ [35]

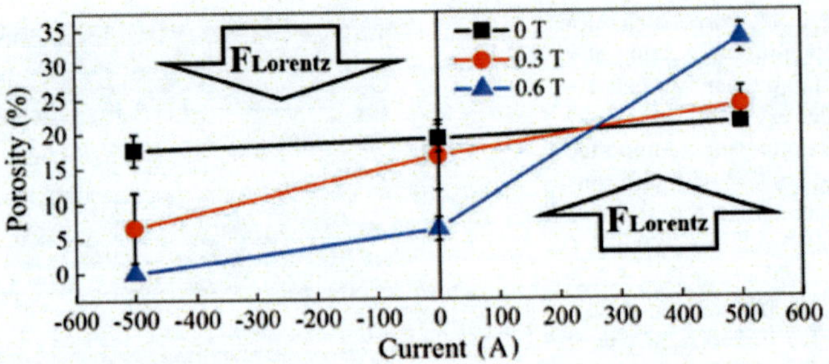

Fig. 3.30 Statistical graphs for the porosity of cladding layers produced under different electromagnetic compound field conditions [35]

$$[C] + [FeO] = Fe + CO \tag{3.42}$$

$$[C] + [MnO] = Mn + CO \tag{3.43}$$

$$2[C] + [SiO_2] = Si + 2CO \tag{3.44}$$

Moreover, the shielding gas can be introduced into the molten pool to a certain extent [39]. Finally, the original depositing powder may contain a certain number of pores [42]. Pores formed by the latter two mechanisms are generally considered uniformly distributed and relatively small. However, the present experimental results show that pores in the cladding layer are extremely large. Therefore, it is believed that most pores in the present study are reactive pores.

Figure 3.31a schematically shows the process of pore formation in the laser cladding layer. With the surface metal inevitably oxidized, oxygen atoms can be transported into the molten pool with the effect of the Marangoni flow, and then react with carbon atoms producing CO gas. Meanwhile, pores that are able to move freely within the molten pool can grow or merge with other pores to form larger pores under Vander Waals forces [43]. As for the detachment process, the surface tension force can hinder a growing pore from breaking away [39]. During the rapid solidification process in laser cladding, pores do not have enough time to escape, thus remain in the cladding layer.

For the condition with the electromagnetic compound field applied, current flows around the pore with no current inside the pore, as the fluid around the pore has high conductivity. When the current is coupled with the magnetic field, it produces the Lorentz volume force, as shown in Fig. 3.31b.

Buoyancy caused by the Lorentz force in the metal melt around the pore is formulated as follows.

$$F_j = \frac{3}{2}\left(\frac{\sigma_p - \sigma_f}{\sigma_p + 2\sigma_f}\right)V_{pore} \times \boldsymbol{J} \times \boldsymbol{B} \tag{3.45}$$

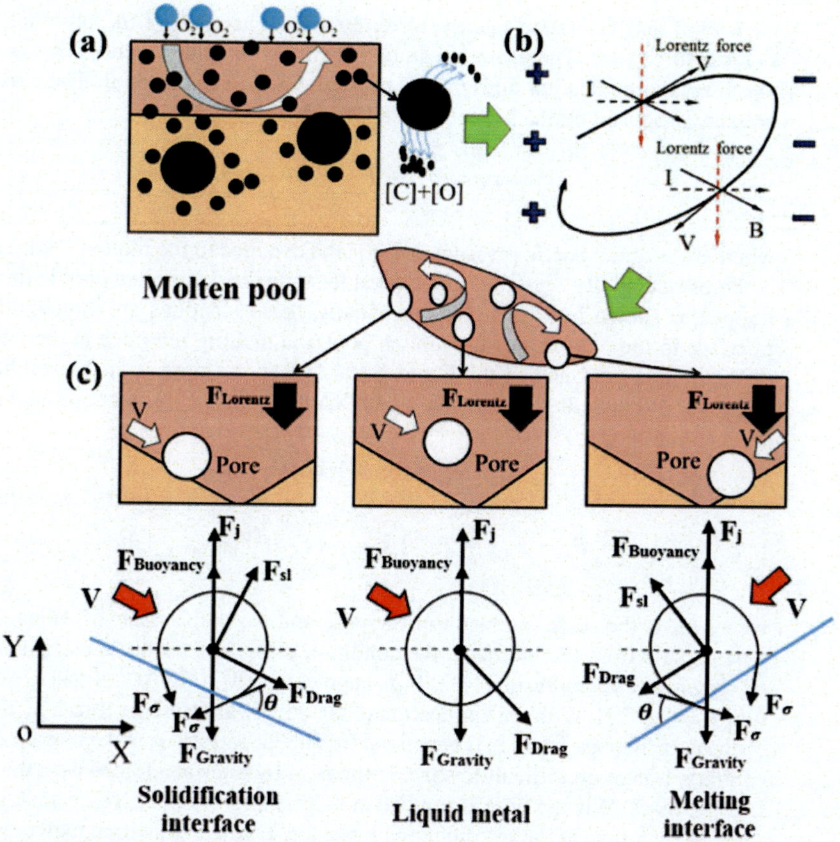

Fig. 3.31 Schematic diagrams of pore reaction mechanisms in molten pool: **a** pore formation process in laser cladding, **b** Lorentz force analysis and **c** pore and force analyses in different positions [35]

where ε_p is the electrical conductivity of the gas phase, ε_f is the electrical conductivity of the molten metal. Since a pore is a gas phase with approximately zero electrical conductivity, Eq. (3.45) becomes

$$F_j = -\frac{3}{4} V_{pore} \times \boldsymbol{J} \times \boldsymbol{B} \tag{3.46}$$

where the negative sign means that the Lorentz force and the corresponding force act on the pore in opposite directions. The directional Lorentz force is in an order of approximately 10^5 N/m^3, whereas the converting pressure produced by the Lorentz force is in an order of 10^2 Pa, which is far less than the atmospheric pressure of 10^5 Pa. It has been reported that the critical pore radius for heterogeneous and homogeneous nucleation and activation energy were almost unchanged by the presence of the Lorentz force [44]. Therefore, it can be concluded that the Lorentz force is unable to inhibit the nucleation and growth of pores in the molten pool. The converting pressure produced by the Lorentz force is given as

$$p_c = p_{at} + F_{Lorentz}h + \rho_l gh \tag{3.47}$$

where p_{at} is atmospheric pressure and h is the distance to the molten pool surface.

Pores potentially occur at three typical locations in the molten pool before being trapped as shown in Fig. 3.31c. Specifically, these locations are the solidification interface in the rear area of the molten pool, the melting interface in the front area of the molten pool and in the liquid metal. When the pore is at the solidification interface, forces acting on the pore in the X-direction and Y-directions are [45]

$$\sum F_x = \mathrm{F}_\sigma \mathrm{x} + \mathrm{F}_{\mathrm{sly}} + \mathrm{F}_{\mathrm{Dragx}} \tag{3.48}$$

$$\sum F_y = \mathrm{F}_\sigma \mathrm{y} + \mathrm{F}_{\mathrm{sly}} + \mathrm{F}_{\mathrm{Dragy}} + F_j + \mathrm{F}_{\mathrm{Buoyancy}} + \mathrm{F}_{\mathrm{Gravity}} \tag{3.49}$$

where F_σ is the surface tension of the pore and F_{sl} is the shear lift force. When a pore departs from the interface, the condition of $\sum F_y > 0$ must be satisfied. Based on previous research results [11], $F_{sl}$is approximately 10^{-8}–10^{-9} N that is much less than $F_j = 10^{-6}$ N. With the application of the downward Lorentz force, $\sum F_y$ acting on the pore increases, which is beneficial for pore departure from the interface. On the contrary, it is more difficult to shackle the pore by interface forces with the upward Lorentz force. When a pore is trapped at the melting interface, force analysis is the same as for a pore at the solidification interface. However, the pore may move to the liquid metal easily because of the melting of the interface. In this case, the melting interface has less influence on pore formation than the solidification interface.

When the pore is in the liquid metal by the forces acting on it as shown in Fig. 3.31c, the following equation with the different Lorentz forces can be established:

$$\sum F_{iy} = \mathrm{F}_{\mathrm{Dragy}} + F_j + \mathrm{F}_{\mathrm{Buoyancy}} + \mathrm{F}_{\mathrm{Gravity}} \tag{3.50}$$

When the Lorentz force acting on the fluid turns to downward, ie. in the same direction with gravity, F_j becomes positive as in Eq. (3.46), resulting in $\sum F_{iy}$ increasing. Meanwhile, laser process parameters (laser power, laser scanning speed and powder feeding rate) are kept constant during the continuous cladding process. Simulation results show that the molten pool solidification rate increases with the action of the downward Lorentz force, as shown in Fig. 3.26. However, as pore movement in the Y-direction is accelerated, a dense coating is formed. Conversely, when there is an upward Lorentz force, it is opposite to gravity, depressing the action of $\sum F_{iy}$ on the pore, as demonstrated by Eq. (3.50). The Y-direction velocity of the pore is suppressed to an extent that the pore can hardly escape from the liquid metal, as displayed in Fig. 3.28a. In the meantime, pores may also merge with each other to form large pores, as revealed by experimental observations shown in Fig. 3.29e.

3.9 Summary of Electromagnetic Field-Assisted Laser Process

An advanced approach to control particle distribution has been proposed for the laser melt injection (LMI) process, which applies an electromagnetic compound field to assist with the fabrication of WC reinforced stainless steel AISI 316L composite coatings. The distributions of fluid temperature, fluid velocity and reinforcement particles in the molten pool were simulated using a 2D multi-physics model coupled with equations of heat transfer, fluid dynamics, drag force, the Lorentz force and phase transition. Results show that the directional Lorentz force, as a type of volume force, can change the equivalent buoyancy acting on particles in the molten pool due to the presence of the electromagnetic compound field. When the Lorentz force and gravity force are in same direction, the majority of particles are trapped in the upper region of the LMI layer. On the other hand, if the Lorentz force and gravity force are in opposite direction, most particles are concentrated in the bottom region of the molten pool. Therefore, the distribution gradient of WC particles can be controlled by the electromagnetic field instead of the time-consuming adjustment of process parameters.

Pores as defects inevitably occur in laser cladding layers. In addition to controlling the distribution gradient of ceramic particles in metal matrix composite coatings, the electromagnetic compound field has also been utilized to improve the escaping ability of pores from the molten pool in laser cladding. Based on the multi-physical field coupling theory, a laser cladding molten pool model to produce nodular cast iron QT-400 was created considering factors such as heat transfer, fluid flow, morphology, buoyancy, surface tension, the Lorentz force and pore movement in the molten pool. The corresponding experiment was carried out in parallel to verify the simulation results. Based on model predictions the fluid flow status and geometry of the molten pool are changed significantly due to the presence of the directional Lorentz force. Pore escape velocity in the molten pool can be improved effectively with the downward Lorentz force generated by a current of 500 A and a steady magnetic flux density of 1.2 T. On the contrary, more pores are trapped inside the molten pool when the direction of the Lorentz force is changed to upward. Therefore, the escape behavior of pores from the molten pool in laser cladding can be effectively controlled. Owing to the downward Lorentz force, dense cladding layers with low porosity can be achieved successfully.

References

1. D. Liu, Y. Chen, L. Li, F. Li, In situ investigation of fracture behavior in monocrystalline WCp-reinforced Ti–6Al–4V metal matrix composites produced by laser melt injection. Scrip. Mater. **59**, 91–94 (2008)
2. D. Miracle, Metal matrix composites—From science to technological significance. Comp. Sci. Technol. **65**, 2526–2540 (2005)
3. W. Tian, L. Qi, C. Su, J. Zhou, Z. Jing, Numerical simulation on elastic properties of short-fiber-reinforced metal matrix composites: Effect of fiber orientation. Comp. Struc. **152**, 408–417 (2016)
4. J.A. Vreeling, V. Ocelík, Y.T. Pei, D.T.L. van Agterveld, J.T.M. De Hosson, Laser melt injection in aluminum alloys: on the role of the oxide skin, Acta Mater. **48**, 4225–4233 (2000)
5. J.A. Vreeling, V. Ocelík, J.T.M. De Hosson, Ti–6Al–4V strengthened by laser melt injection of WCp particles. Acta Mater. **50**, 4913–4924 (2002)
6. Y.T. Pei, V. Ocelik, J.T.M. De Hosson, SiCp/Ti6Al4V functionally graded materials produced by laser melt injection. Acta Mater. **50**, 2035–2051 (2002)
7. D. Liu, L. Li, F. Li, Y. Chen, WCp/Fe metal matrix composites produced by laser melt injection. Surf. Coat. Technol. **202**, 1771–1777 (2008)
8. A.M. Do Nascimento, V. Ocelík, M.C.F. Ierardi, J.T.M. De Hosson, Microstructure of reaction zone in WCp/duplex stainless steels matrix composites processing by laser melt injection, Surf. Coat. Technol. **202**, 2113–2120 (2008)
9. L. Wang, J.H. Yao, H. Yong, Q.L. Zhang, Z. Sun, R. Liu, Influence of electric-magnetic compound field on the WC particles distribution in laser melt injection. Surf. Coat. Technol. **315**, 32–43 (2017)
10. M. Gatzen, Z. Tang, F. Vollertsen, Effect of electromagnetic stirring on the element distribution in laser beam welding of aluminium with filler wire. Phys. Procedia **12**, 56–65 (2011)
11. M. Gatzen, Influence of low-frequency magnetic fields during laser beam welding of aluminium with filler wire. Phys. Procedia **39**, 59–66 (2012)
12. S. Yin, X. Wang, X. Suo, H. Liao, Z. Guo, W. Li, C. Coddet, Deposition behavior of thermally softened copper particles in cold spraying. Acta Mater. **61**, 5105–5118 (2013)
13. S.Z. Shuja, B.S. Yilbas, H. Ali, C. Karatas, Laser pulse heating of steel mixing with WC particles in a irradiated region. Opt. Laser Technol. **86**, 126–135 (2016)
14. A. Prosperetti, G. Tryggvason, *Computational Methods for Multiphase Flow* (Cambridge University Press, Cambridge, 2009), pp. 1–18
15. G. Kaptay, Interfacial criterion of spontaneous and forced engulfment of reinforcing particles by an advancing solid/liquid interface. Metall. Mater. Trans. A **32**, 993–1005 (2001)
16. I. Tabernero, A. Lamikiz, E. Ukar, L.N. López de Lacalle, C. Angulo, G. Urbikain, Numerical simulation and experimental validation of powder flux distribution in coaxial laser cladding. J. Mater. Process. Technol. **210**, 2125–2134 (2010)
17. S. Morville, M. Carin, P. Peyre, M. Gharbi, D. Carron, P. Le Masson, R. Fabbro, 2D longitudinal modeling of heat transfer and fluid flow during multilayered direct laser metal deposition process. J. Laser Appl. **24**, 032008 (2012)
18. A.D. Brent, V.R. Voller, K.J. Reid, Enthalpy-porosity technique for modeling convection-diffusion phase change: Application to the melting of a pure metal. Numer. Heat Trans. **13**(3), 297–318 (1988)
19. Y. Fu, A. Loredo, B. Martin, A.B. Vannes, A theoretical model for laser and powder particles interaction during laser cladding. J. Mater. Process. Technol. **128**, 106–112 (2002)
20. O. Schenk, K. Gartner, W. Fichtner, A. Stricker, PARDISO: a high-performance serial and parallel sparse linear solver in semiconductor device simulation. Fut. Gener. Comp. Sys. **18**(1), 69–78 (2001)
21. T. Amine, J.W. Newkirk, F. Liou, Investigation of effect of process parameters on multilayer builds by direct metal deposition. Appl. Therm. Eng. **73**, 500–511 (2014)

22. J. Rahman Chukkan, M. Vasudevan, S. Muthukumaran, R. Ravi Kumar, N. Chandrasekhar, Simulation of laser butt welding of AISI 316L stainless steel sheet using various heat sources and experimental validation, J. Mater. Process. Technol. **219**, 48–59 (2015)
23. B.H. Dennis, G.S. Dulikravich, Magnetic field suppression of melt flow in crystal growth. Int. J. Heat Fluid Flow **23**, 269–277 (2002)
24. D. Shangguan, S. Ahuja, D.M. Stefanescu, An analytical model for the interaction between an insoluble particle and an advancing solid/liquid interface. Metall. Mater. Trans. A **23A**, 669–680 (1992)
25. F. Arias-González, J. del Val, R. Comesaña, J. Penide, F. Lusquiños, F. Quintero, A. Riveiro, M. Boutinguiza, J. Pou, Fiber laser cladding of nickel-based alloy on cast iron. Appl. Surf. Sci. **374**, 197–205 (2016)
26. C.L. Zhong, A. Gasser, J. Kittel, K. Wissenbach, R. Poprawe, Improvement of material performance of Inconel 718 formed by high deposition-rate laser metal deposition. Mater. Des. **98**, 128–134 (2016)
27. C. Zhou, S. Zhao, Y. Wang, F.L. Liu, W.Y. Gao, X.C. Lin, Mitigation of pores generation at overlapping zone during laser cladding. J. Mater. Process. Technol. **216**, 369–374 (2015)
28. I. Tabernero, A. Lamikiz, S. Martinez, E. Ukar, J. Figueras, Evaluation of the mechanical properties of Inconel 718 components built by laser cladding. Int. J. Mach. Tool Manuf. **51**, 465–470 (2011)
29. Z.K. Weng, A.H. Wang, Y.Y. Wang, D.H. Xiong, H.Q. Tang, Diode laser cladding of Fe-based alloy on ductile cast iron and related interfacial behavior. Surf. Coat. Technol. **286**, 64–71 (2016)
30. O. Velde, A. Techel, R. Grundmann, Suppression of the development of pores during laser-induced surface dispersion of TiC into aluminium by means of a static magnetic field. Surf. Coat. Technol. **150**, 170–176 (2002)
31. M. Bachmann, V. Avilov, A. Gumenyuk, M. Rethmeier, Experimental and numerical investigation of an electromagnetic weld pool support system for high power laser beam welding of austenitic stainless steel. J. Manuf. Process. **214**, 578–591 (2014)
32. N. Zhang, W.W. Liu, D.W. Deng, Z.J. Tang, X.Y. Liu, Z.R. Yan, H.C. Zhang, Effect of electric-magnetic compound field on the pore distribution in laser cladding process. Opt. Laser Technol. **108**, 247–254 (2018)
33. L. Chen, Y. Yang, F.L. Jiang, C. Li, Experimental investigation and FEM analysis of laser cladding assisted by coupled field of electric and magnetic. Mater. Res. Express **6**, 016516 (2018)
34. J.Z. Zhou, J.L. Xu, S. Huang, Z.R. Hu, X.K. Meng, X. Fneg, Effect of laser surface melting with alternating magnetic field on wear and corrosion resistance of magnesium alloy. Surf. Coat. Technol. **309**, 212–219 (2017)
35. Y. Hu, L. Wang, J.H. Yao, H.C. Xia, J.H. Li, R. Liu, Effects of electromagnetic compound field on the escape behavior of pores in molten pool during laser cladding. Surf. Coat. Technol. **383**, 125198 (2020)
36. Z.T. Gan, H. Liu, S.X. Li, X.L. He, G. Yu, Modeling of thermal behavior and mass transport in multi-layer laser additive manufacturing of Ni-based alloy on cast iron. Int. J. Heat Mass Transfer. **111**, 709–722 (2017)
37. Z.Y. Liu, H. Qi, Numerical simulation of transport phenomena for a double-layer laser powder deposition of single-crystal superalloy. Metall. Mater. Trans. A-Phys. Metall. Mater. Sci. **45A**, 1903–1915 (2014)
38. S.Y. Wen, Y.C. Shin, Modeling of transport phenomena in direct laser deposition of metal matrix composite. Int. J. Heat Mass Transfer. **54**, 5319–5326 (2011)
39. Z.X. Yang, A.H. Wang, D.H. Xiong, B. Ye, X.Y. Qi, Porosity elimination and heat treatment of diode laser-clad homogeneous coating on cast aluminum-copper alloy. Surf. Coat. Technol. **321**, 26–35 (2017)
40. J.F. Klausner, R. Mei, D.M. Bernhard, L.Z. Zeng, Vapor bubble departure in forced convection boiling. Int. J. Heat Mass Transfer. **36**, 651–662 (1993)

41. R. Mei, J.F. Klausner, Unsteady force on a spherical bubble at finite Reynolds number with small fluctuations in the free-stream velocity. Phys. Fluid. **4**, 63–70 (1992)
42. C.L. Zhong, A. Gasser, T. Schopphoven, R. Poprawe, Experimental study of porosity reduction in high deposition-rate Laser Material Deposition. Opt. Laser Technol. **75**, 87–92 (2017)
43. Y.T. Tao, A. Yeckel, J.J. Derby, Steady-state and dynamic models for particle engulfment during solidification. J. Comput. Phys. **315**, 238–263 (2016)
44. H.W. Zhang, Study on bubble nucleation in liquid metal. Acta Phys. Sin. **56**, 4864–4871 (2007)
45. D. Leenov, A. Kolin, Theory of electromagnetophoresis. I. magnetohydrodynamic forces experienced by spherical and symmetrically oriented cylindrical particles. J, Chem. Phys. **22**, 683–688 (1954)

Chapter 4
Supersonic Laser Deposition of Metals

Abstract Cold spray (CS) is a relatively new materials deposition technique discovered in last thirty years and has been rapidly developing during past two decades, owing to many unique features. In the CS process, deleterious effects such as high temperature oxidation, evaporation, melting, residual stresses and other common problems inherent to traditional thermal spraying (such as flame spraying, arc spraying, plasma spraying, etc.) can be minimized or eliminated. Certainly, preheating of the deposited particles and substrate can favor the adhesion of the coating to the substrate, in particular, when depositing hard materials. Therefore, it would be desirable that introducing additional heat sources into CS can thermally soften the deposited particles and/or substrate. Laser, as one of the most ideal heat sources for material processing due to its unique advantages such as high energy density, chemically clean and flexible operation, has been coupled with CS. This combined process is termed as supersonic laser deposition (SLD). This novel coating fabrication technique has been employed to deposit both soft materials, for example, copper, and hard materials such as Stellite alloys, Ni60 alloy, etc. The experimental results have demonstrated that SLD surpasses CS to produce a dense coating with good interfacial bonding between particles and between particles and substrate for both soft and hard materials. The coatings prepared via SLD are compared with those by cold spray or laser cladding, showing that SLD can minimize oxidation of deposited materials, so the prepared coatings contain much fewer defects, simultaneously, and display strong interface bonding by both mechanical interlocking and metallurgical diffusion mechanisms. This novel deposition technique utilizing laser provides the coating industry an effective means to deposit hard materials which are temperature-sensitive.

4.1 Supersonic Laser Deposition

4.1.1 Cold Spray

Cold spray (CS) is a relatively new material deposition technique whereby metal powders ranging in particle size from 5 to 100 μm are accelerated to a high velocity

J. Yao et al., *Advanced Laser Process for Surface Enhancement*, Advanced Topics in Science and Technology in China 61, https://doi.org/10.1007/978-981-15-9659-9_4

in a supersonic gas flow and then impinged onto the substrate or already deposited coating at a temperature well below the melting point of sprayed materials [1]. Intensive plastic deformation induced by high-velocity impact occurs in solid state particle, substrate (or already deposited coating) or both, enabling formation of a less-oxidized cold-sprayed coating [2]. In the past few years, a wide range of pure metals, metallic alloys, polymers and composites have been successfully deposited onto a variety of substrate materials using CS [3]. One of the most widely used concepts in CS is critical velocity which is material-dependent [4]. Small particles easier achieve high velocities than large particles. Since powders generally contain a mixture of particles with various sizes, some fraction of the powder is deposited on the substrate while the remainder bounces off. The ratio of the weight of powder deposited onto the substrate to that of the total powder used is called deposition efficiency. It is evident that this ratio can be increased by reducing critical velocity and/or increasing impact velocity.

It has been suggested that an increase in particle and/or substrate temperature can raise deposition efficiency and also reduce critical deposition velocity as a result of softening effect on the materials [5]. Moreover, increased substrate temperature may also help promote the interface bonding of the coatings, since CS coatings generally have lower bonding strength than those produced by other deposition methods because of the short interaction time between particles and substrate [6]. However, it has also been noted that elevating particle temperature by raising gas temperature increases the risk of nozzle fouling when spraying low melting point metals. As a result, a method of simultaneous heating the particles and substrate without clogging the Laval nozzle would be desired [7].

4.1.2 *Laser Aided Cold Spray*

In recent years, a novel deposition technique, known as supersonic laser deposition (SLD), has been developed by introducing laser into CS process [8]. In SLD, the deposition site of CS is simultaneously heated by laser in order to preheat the particles and soften the substrate as well. This method of heating the particles not only avoids nozzle fouling but also significantly reduces the critical velocity, which allows bonding to occur upon impact at the velocity about half that in CS. Because of these unique features, SLD technology has been applied on various materials, in particular, hard materials such as Stellite alloys, Ni60, diamond, and so on [9–18].

4.2 Supersonic Laser Deposition of Copper

4.2.1 Copper Coating Deposition

Copper, as a coating material, can be applied on filler wires of welding for corrosion protection. It is also used on implants as an alternative antimicrobial coating. In the production of stainless steel wire, it is common to coat the stainless steel stock with metallic copper, which serves as a wiredrawing lubricant for the stock passing through the dies. In the present research group, commercially available pure Cu powder (99.72%) was deposited on carbon steel with SLD [9]. The Cu powder generally has irregular shape with the size about 25 μm on average, as shown in Fig. 4.1, which was determined by a laser diffraction particle size analyzer (Horiba LA-950, Japan). The substrate material is medium-carbon steel with the chemical composition (wt.%) consisting of 0.43% C, 0.23% Si, 0.66% Mn, Bal. Fe. The specimen has a dimension of 100 × 60 × 10 mm with thickness of 10 mm. Before the deposition process, the substrate surface was grit-blasted using 24 mesh alumina and ultrasonic cleaned in alcohol.

The schematic diagram of the SLD system used in this study is shown in Fig. 4.2, which includes cold spray system, laser unit, high pressure gas supply unit and other auxiliary equipment such as robot. High pressure air about 30 bar was supplied to a convergingdiverging nozzle in two directions; one was through the gas heater, the other was via a powder feeder where Cu powder was held. The Cu feedstock stream

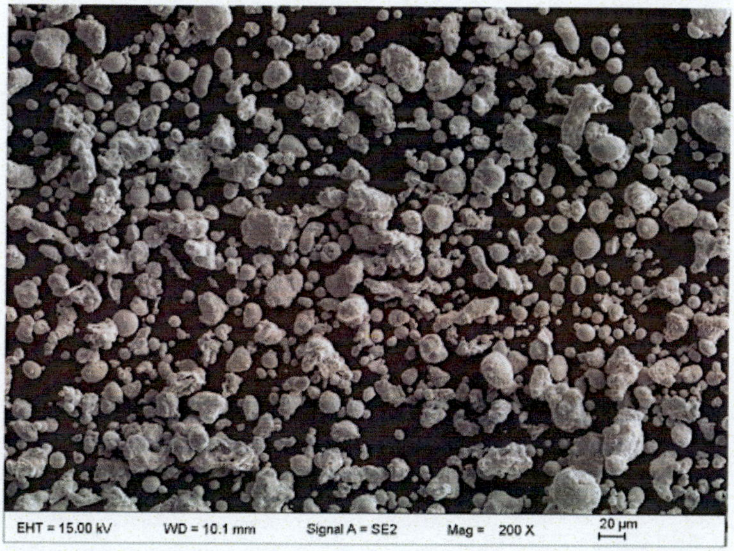

Fig. 4.1 SEM morphology of Cu powder [9]

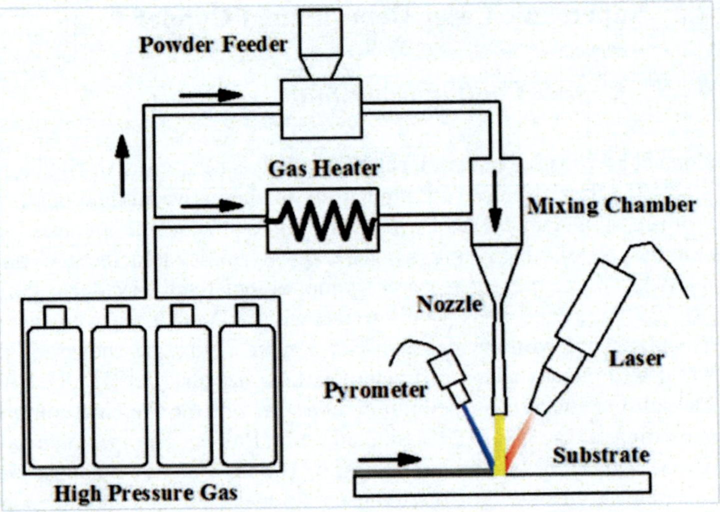

(a)

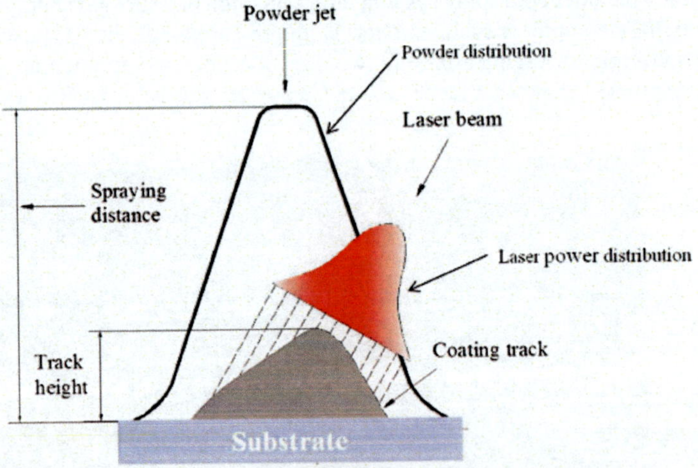

(b)

Fig. 4.2 Schematic diagram of **a** SLD system and **b** laser beam and coating track configuration in SLD process [9] and air were mixed and passed through the nozzle where the Cu particles were accelerated to supersonic speed. The high-velocity particles impacted an area of the substrate which was synchronously heated by a diode laser with 960–980 nm wavelengths and 4 kW maximum power (Laserline LDF 400–1000, Germany). Combined lenses were used to focus the laser beam with the spot diameter of 5 mm onto the substrate surface. Laser power was controlled by a feedback system which utilized an infra-red pyrometer with the temperature control precision of ±2 °C to monitor deposition zone temperature. The nozzle, laser head and pyrometer were installed on a robot (STAÜBLI TX 90, Switzerland). The cold spray system was designed and made by the Chinese Academy of Sciences. An in-house made de Laval nozzle with an area expansion ratio of ~8.3 and a length of 180 mm was employed.

In the deposition process, the substrate was stationary and the nozzle, laser head and pyrometer were moveable, controlled by the robot. As illustrated in Fig. 4.2, the spraying nozzle was perpendicular to the substrate surface and the laser beam was at an angle (30°) to the surface normal. The powder stream and the laser beam partially overlapped with each other. Therefore, during the deposition process the impinging particles were also partially irradiated by the laser beam prior to impacting the substrate. The working gas was air. The Cu coating was deposited under a wide range of the process parameters that were gas pressure, gas temperature, laser power, laser and robot traverse speed, powder feeding rate and standoff distance, in order to obtain optimal deposition condition with respect to best coating surface morphology and interface bonding, maximum coating thickness and density, minimum coating oxidation and cracking, highest hardness. The obtained optimal process parameters in such a way were gas pressure of 2.4 MPa, gas temperature of 400 °C, laser power of 800 W, laser traverse speed of 10 mm/s, powder feeding rate of 40 g/min, standoff distance of 30 mm. In the SLD process, the laser power was elaborately controlled to only soften the spraying particles and substrate surface, but not to melt them. For comparison the CS Cu coating was also prepared under the same condition except that laser irradiation was not involved.

4.2.2 Copper Coating Morphology

The coating surface roughness was characterized using a three-dimensional laser microscopy (LEXT OLS4100, OLYMPUS) and the surface profiles are shown in Fig. 4.3. It can be seen that the surface roughness of the coating prepared by CS appears higher than that of the coating prepared by SLD. This suggests that laser irradiation can improve the surface morphology of the Cu coating. The surface roughness was quantified to be Ra 27.6 μm for the CS coating and Ra 22.2 μm for the SLD coating.

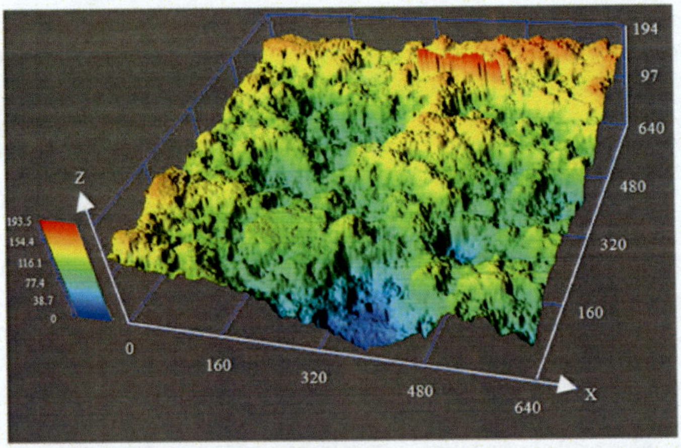

(a)

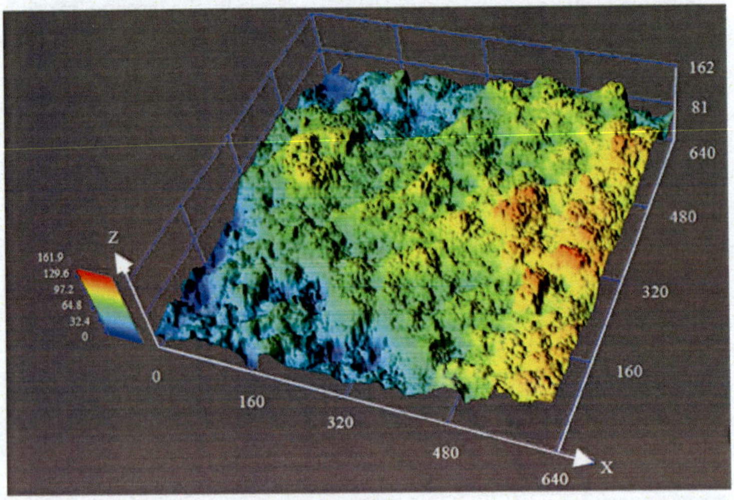

(b)

Fig. 4.3 Three-dimensional surface roughness profile of **a** CS coating and **b** SLD coating [9]

The comparison of coating thickness between the CS and SLD coating is illustrated in Fig. 4.4. The SLD coating is thicker than the CS coating. The peak coating thickness of the CS coating is about 1.3 mm, while that of the SLD coating is about 2.2 mm. Laser irradiation increased the peak coating thickness by about 70%. In other words, laser irradiation improved the deposition efficiency significantly, which is attributed to preheating of the particles and substrate. Furthermore, deposition efficiency (DE), which is defined as the weight of successfully bonded particles onto

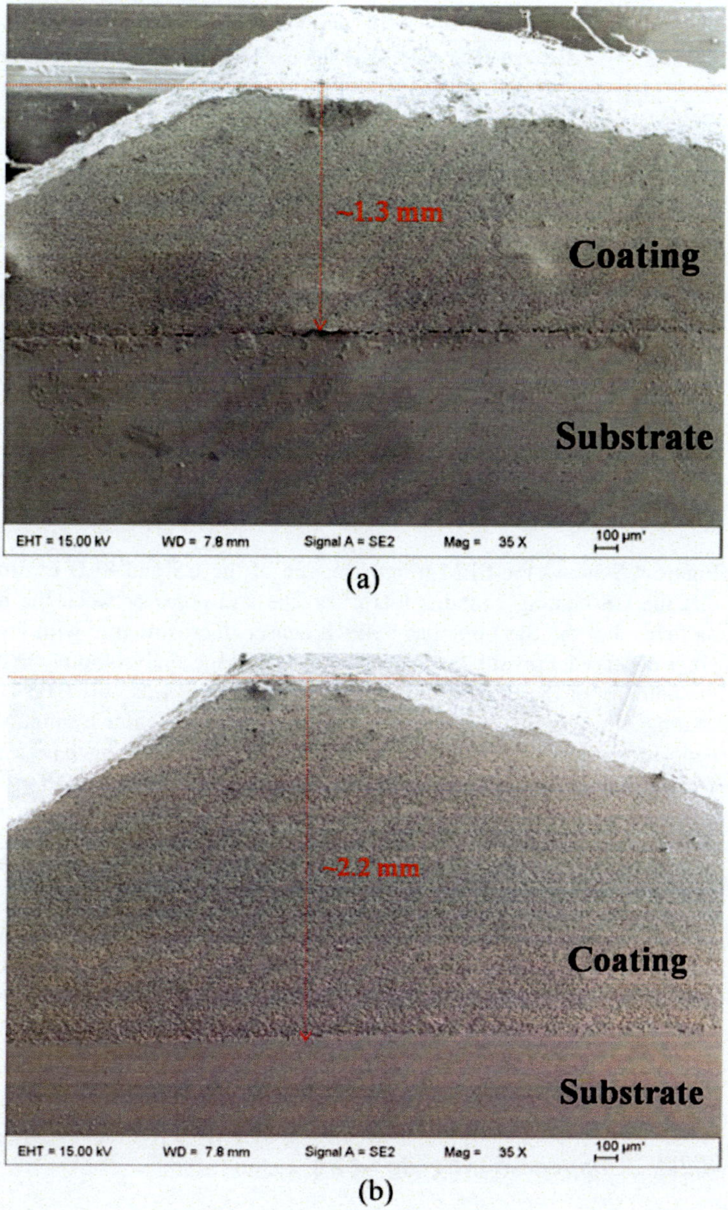

Fig. 4.4 Coating thickness of **a** CS coating and **b** SLD coating [9]

Table 4.1 Weight change of substrate and total weight of feedstock powder [9]

Coating	Substrate weight before deposition (g)	Substrate weight after deposition (g)	Total weight of feedstock powder (g)
CS coating	171.42	182.30	40
SLD coating	174.19	195.33	40

the substrate divided by the total weight of initial feedstock powder used [19]. The experimental data of the weight change of the substrate before and after deposition and the total weight of initial feedstock powder used are summarized in Table 4.1. Thus, deposition efficiency was calculated to be 27.2% for the CS coating and 52.9% for the SLD coating. The deposition efficiency of the CS coating was improved by about two times with the assistance of laser irradiation.

4.2.3 Copper Coating Microstructure

Figure 4.5 shows the SEM microstructure of the CS and SLD coating. It is seen that the CS coating contains lots of cracks and pores between the deformed Cu particles, but the SLD one has a much denser microstructure with few cracks and pores observed. Porosity measurements using image analysis software indicated that the porosity of the CS coating was 3.367%, while it was only 0.08% for the SLD coating. This, again, confirms the beneficial effect of laser irradiation on coating deposition in the CS process. Furthermore, it is also interesting to notice that particle deformation in the SLD process is more severe than that in the CS process, as seen in Fig. 4.5. This is mainly due to softening of the particles by laser irradiation. In the CS process, the initially deposited particles are hammered by successive high velocity impacting particles. The softened particles by laser heating are easier to deform by the impact of particles at a high velocity, leading to tightly bonding to the substrate and/or already deposited particles. From the SEM images of cross-section of the coatings in Fig. 4.6, obvious cracks were observed at the interface between the coating layer and substrate of the CS coating, but there were not found in the SLD coating, instead, material penetration occurred at the interface, which enhanced the coating bonding to the substrate.

4.2.4 Copper Coating Bonding

To further investigate the interface bonding strength, adhesion test was performed on the CS and SLD coating. The coating specimen was glued to the loading fixture with an epoxy resin adhesive before tension, it was then subjected to tensile loading which was applied normal to the coating surface until the coating failed at the interface. As

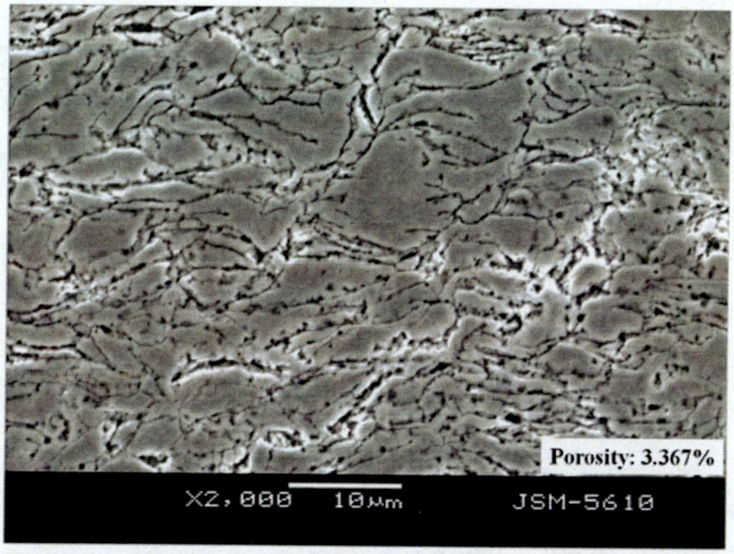

(a)

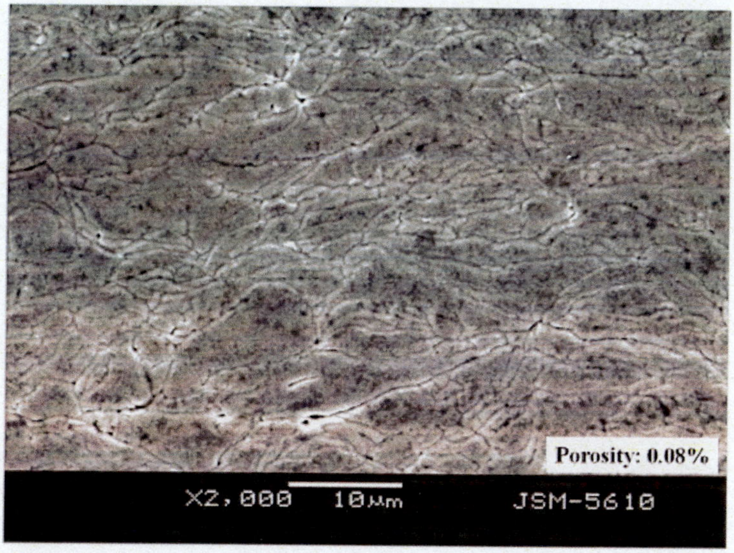

(b)

Fig. 4.5 SEM image of microstructure of **a** CS coating and **b** SLD coating [9]

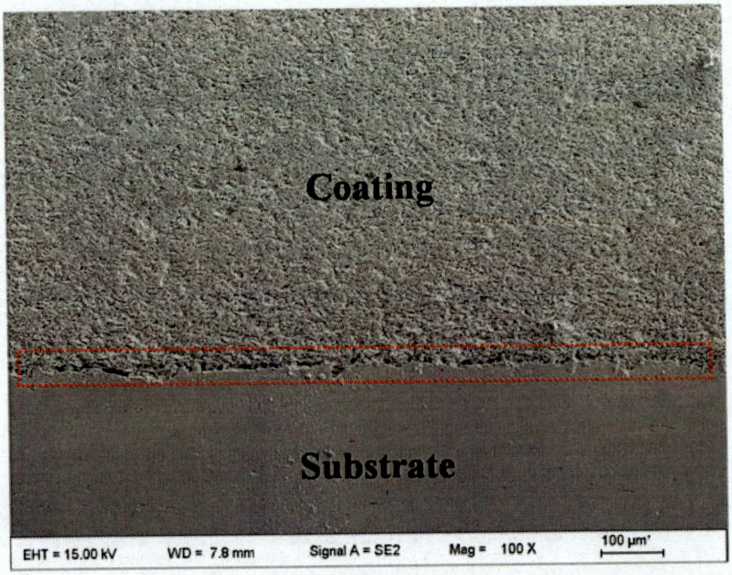

(a)

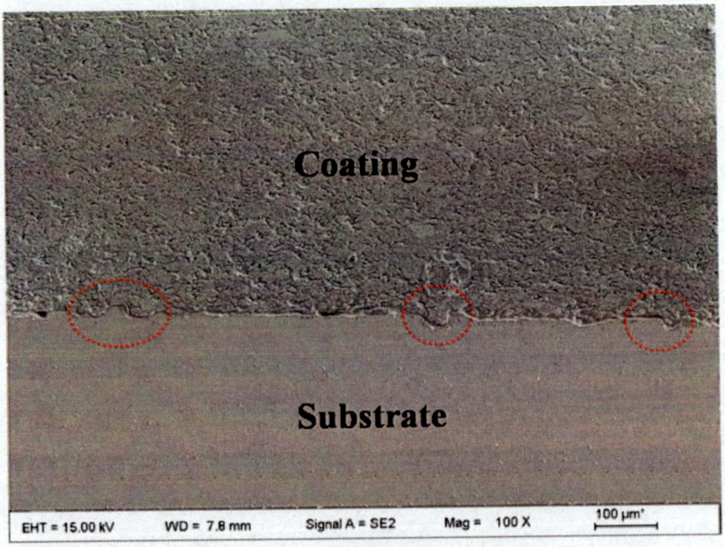

(b)

Fig. 4.6 SEM image in cross-section of **a** CS coating and **b** SLD coating [9]

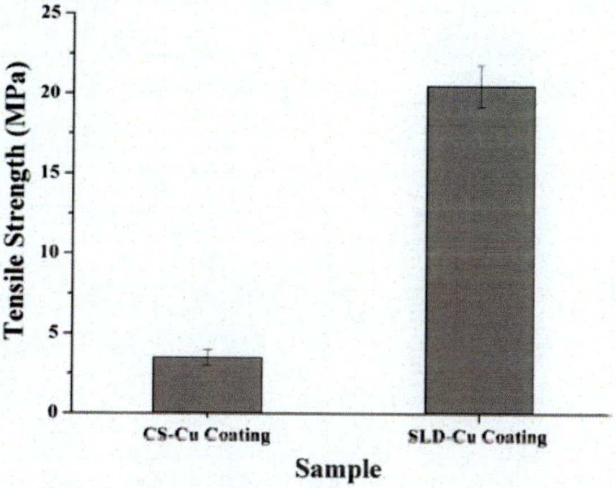

Fig. 4.7 Comparison of adhesion strength between the Cu coatings [9]

illustrated in Fig. 4.7, the adhesion strength of the CS coating is much higher than that of the CS coating, which is consistent with the SEM examination (Fig. 4.6).

The most widely accepted bonding mechanism of CS is the occurrence of adiabatic shear instability at the interface, which results from the high strain rate and the intensive localized deformation of the particles and substrate [20]. The interfacial instability occurring during the high velocity impacting process results in material roll-ups and vortices at the interface. As a consequence, substrate and coating materials merge at the interface region, leading to mechanical interlocking [21]. Another proposed bonding mechanism of CS is the metallurgical bonding by atomic diffusion between coating and substrate materials, which can provide better bonding strength compared to mechanical interlocking [22]. However, in CS the solid-state diffusion is not sufficient to form a thick diffusion layer due to extremely short timescale of the particlesubstrate interaction and low temperature. In the case of synchronous laser irradiation on the deposition site, the substrate is heated up and it is thereby softened. This makes it easier to lodge the particles to form mechanical interlocking. Moreover, the increased substrate temperature can promote atomic diffusion between the coating and substrate materials, which greatly increases the possibility of metallurgical bonding. All these contribute to good interfacial bonding of the SLD coating.

4.2.5 Copper Coating Hardness

The microhardness of the Cu coatings deposited with different laser powers was determined using a Vickers Indenter (HR-150DT, DHT). As illustrated in Fig. 4.8, the hardness of the CS coating appears the highest (140 $HV_{0.01}$) among the coatings. This is attributed to the strain-hardening effect from the high velocity impact of particles in the CS process. However, with synchronous laser irradiation, the hardness

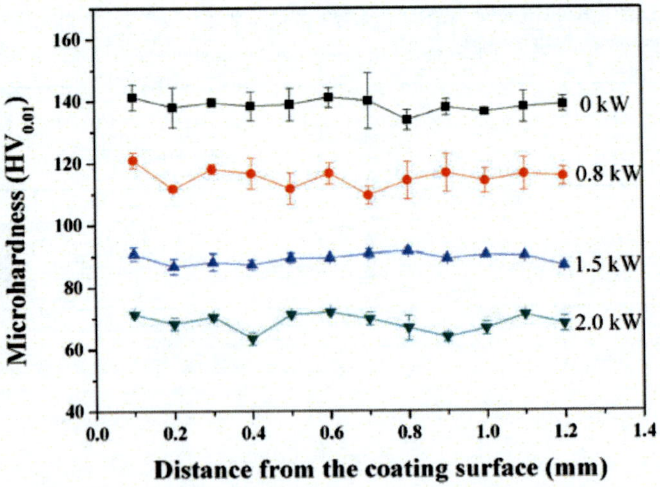

Fig. 4.8 Hardness variation along coating depth [9]

of the deposited coating layer decreased gradually with increasing laser power. It can be seen that the hardness of the SLD coating prepared with 2.0 kW laser power is only about 70 $HV_{0.01}$, which is half of that of the CS coating. Therefore, it can be concluded that synchronous laser irradiation has a softening effect on the deposited coating layer, which is equivalent to annealing treatment of CS coatings. On the other hand, as reported in literature [23], annealing treatment can improve the cohesion strength between particles of CS coatings by healing up the interfacial gaps. From the present study, it can be concluded that SLD as a novel deposition technique surpasses conventional CS method to produce dense coatings with good adhesion to the substrate, but for copper, the coating hardness was reduced due to laser irradiation in the CS process.

4.3 Supersonic Laser Deposition of Stellite Alloy

4.3.1 Stellite 6 Coating Deposition

Compared with copper, Stellite alloys are much harder [24]. As the most popular among Stellite alloy family, Stellite 6 has a wide range of application, one of which is surface coating for wear resistance. Owing to less oxidation influence, cold spray (CS) process is a better means to deposit hard Stellite 6 powder onto various substrates. However, due to insufficient bonding strength of the CS coating, it was proposed to incorporate laser irradiation in the CS deposition of Stellite 6 [10]. Commercially available Stellite 6 powder was used as feedstock material. The morphology and microstructure of the Stellite 6 powder were examined with SEM (ΣIGMA HV-01-043, Carl Zeiss). The particle size distribution of the Stellite 6 powder was investigated using a laser diffraction particle size analyzer (Horiba LA-950, Japan). As shown in Fig. 4.9, the powder particles have a spherical morphology and a typical dendrite microstructure, with an average size about 15 μm. The substrate material is medium-carbon steel with the chemical composition (wt.%) of 0.43% C, 0.23% Si, 0.66% Mn, Bal. Fe. The specimen is a plate having a dimension of 100 × 60 × 10 mm. Before coating process, the substrate surface was grit-blasted using 24 mesh alumina and ultrasonic cleaned in alcohol.

The SLD system used in this study is illustrated in Fig. 4.2. The spatial distribution of laser energy is non-uniform with a configuration like a near top hat intensity distribution, as shown in Fig. 4.10. A high-speed infrared pyrometer was used to obtain real-time temperature measurement and control the temperature of the deposition site, termed as deposition temperature, during the supersonic laser deposition (SLD) process. The data from the pyrometer was fed through a closed-loop feedback system which altered laser power as necessary to maintain the desired temperature. The highest laser power employed in the SLD was measured to be 1.5 kW. The spraying nozzle was perpendicular to the substrate surface and the laser beam was at an angle of 30° to the surface normal. In the deposition process, the substrate was stationary and the nozzle, laser head and pyrometer were moveable, controlled by a robot.

Stellite 6 coatings were deposited onto medium-carbon steel using both laser cladding (LC) and SLD system for comparison. The LC process parameters are laser power of 1.8 kW, laser traverse speed of 8 mm/s, and powder feeding rate of 15 g/min. The overlapping rate between neighboring tracks of each coating is 50%.

4.3.2 Stellite 6 Coating Surface Morphology

The coating surface profile was characterized by a three-dimensional laser microscopy (LEXT OLS4100, OLYMPUS) and the 3D profiles of the SLD coatings prepared with different deposition temperatures are shown in Fig. 4.11. To facilitate the comparison, the average surface roughness of the SLD Stellite 6 against the deposition temperature is plotted in Fig. 4.12. It can be seen that the coating surface prepared with lower deposition temperature is more rugged than that prepared with higher deposition temperature. With increasing the deposition temperature, the surface roughness of the coating is reduced. These results suggest that with the assistance of laser irradiation a smoother coating surface can be achieved, which is beneficial for its post-machining.

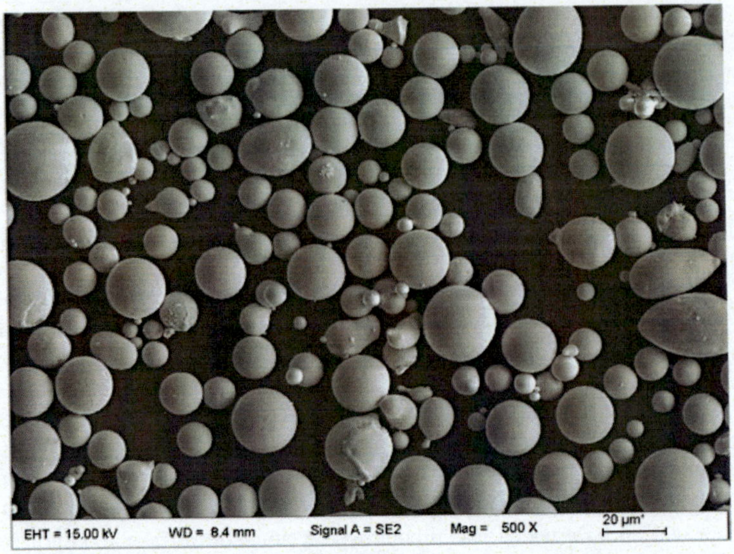

(a)

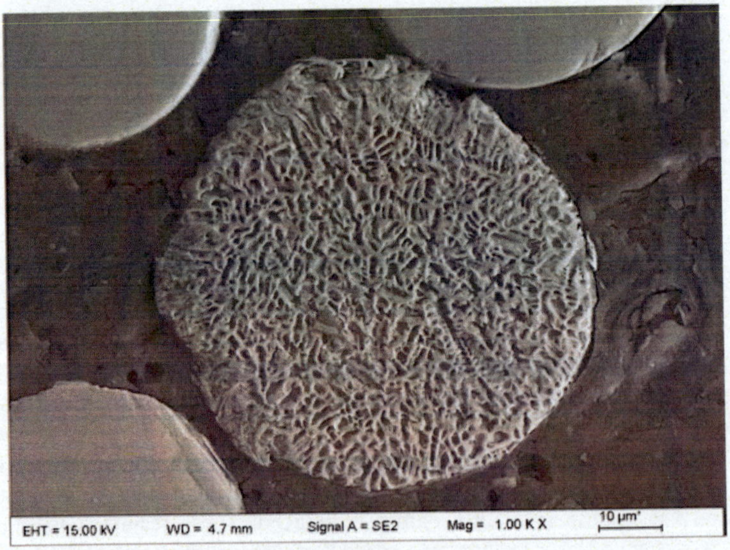

(b)

Fig. 4.9 SEM image of Stellite 6 powder: **a** morphology and **b** particle interior microstructure [10]

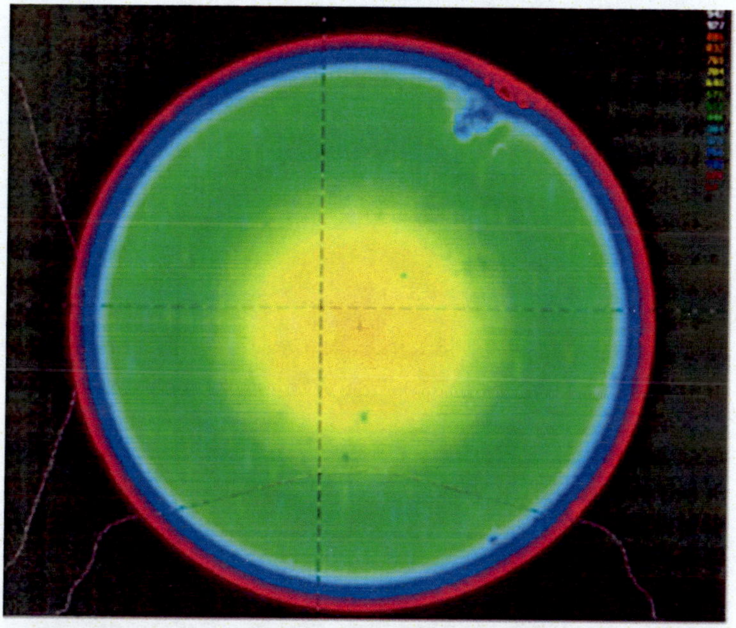

Fig. 4.10 Energy or power distribution in the cross-section of the laser beam in the SLD system [10]

4.3.3 Stellite 6 Coating Thickness

The cross-sections of the SLD Stellite 6 coatings are examined using optical microscopy and the images are shown in Fig. 4.13 for the coatings prepared under different deposition temperatures from 1210 to 1290 °C. It can be seen that all coatings are dense and no cracks are observed. Moreover, the peak coating thickness increases with deposition temperature, which indicates that deposition efficiency can be enhanced with deposition temperature increased. Weighing method was used to quantify the deposition efficiency. First the deposition rate is calculated as

$$R_D = \frac{\Delta m}{t} \tag{4.1}$$

where Δm is the weight change of the substrate after a deposition time t. Since the powder feeding rate R_F was kept constant during the SLD process, the deposition efficiency E can be obtained

$$E = \frac{R_D}{R_F} \times 100\% \tag{4.2}$$

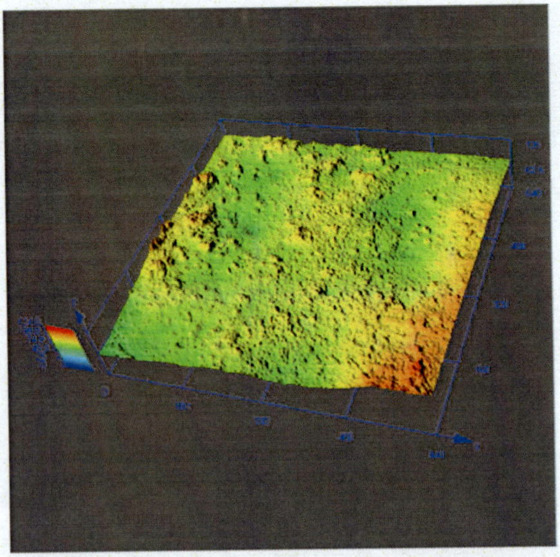

(a)

(b)

Fig. 4.11 3D surface profile of SLD Stellite 6 coating prepared at deposition temperature of **a** 1210 °C, **b** 1230 °C, **c** 1250 °C, **d** 1270 °C and **e** 1290 °C [10]

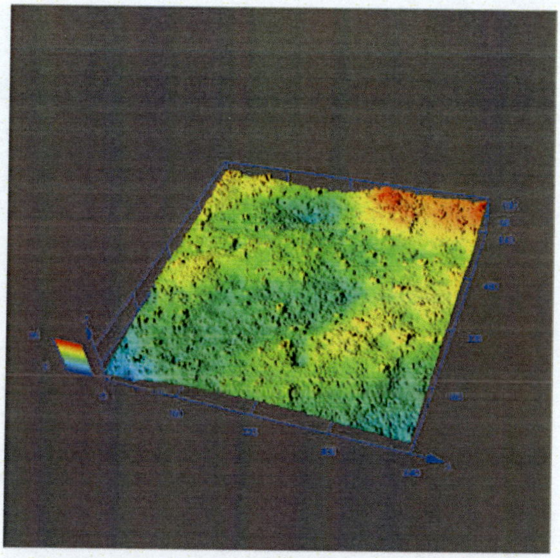

(c)

(d)

Fig. 4.11 (continued)

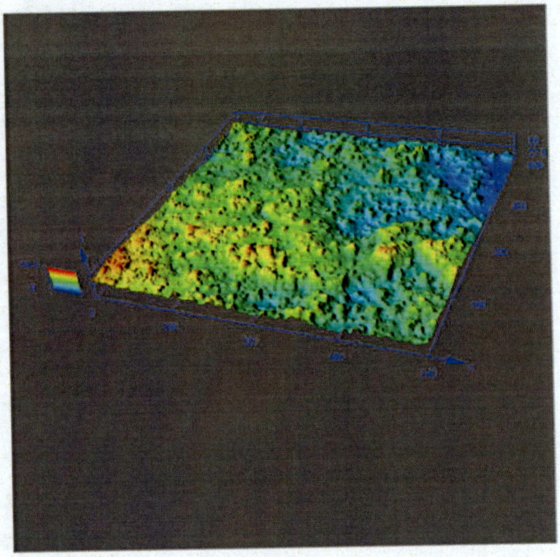

(e)

Fig. 4.11 (continued)

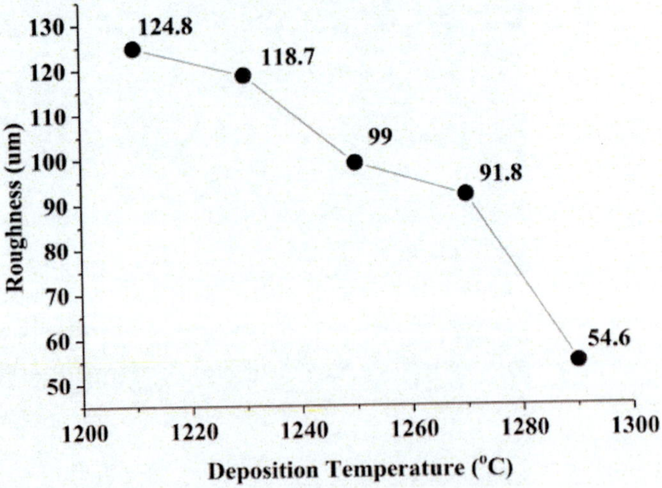

Fig. 4.12 Variation of average surface roughness of SLD Stellite 6 coating with deposition temperature [10]

Fig. 4.13 Optical microscopic image of cross-section of SLD Stellite 6 coating prepared under the deposition temperature of **a** 1210 °C, **b** 1230 °C, **c** 1250 °C, **d** 1270 °C and **e** 1290 °C [10]

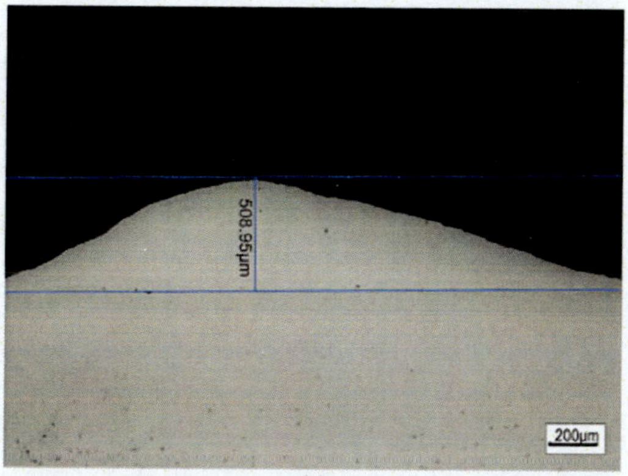

(a)

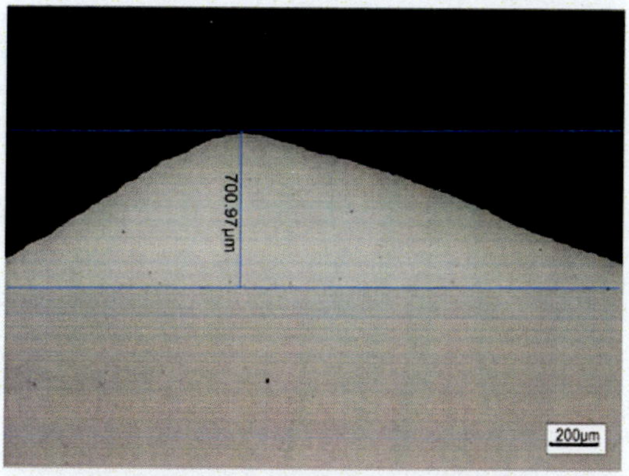

(b)

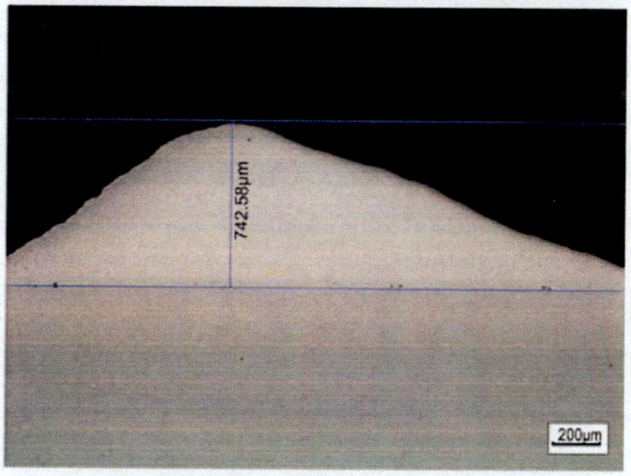

(c)

Fig. 4.13 (continued)

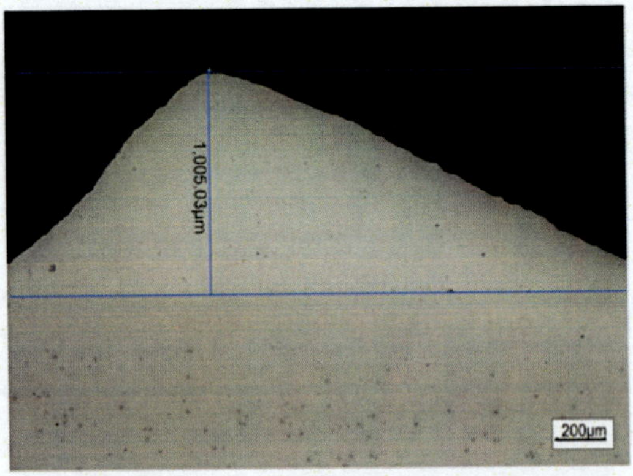

(d)

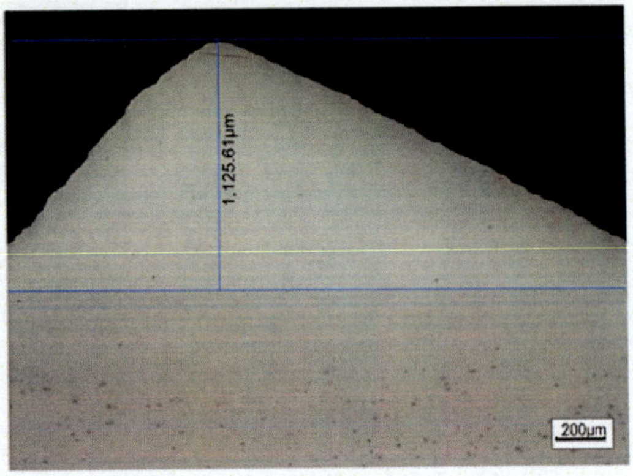

(e)

The deposition efficiency results are plotted in Fig. 4.14 as a function of deposition temperature. As shown, the deposition efficiency increases with deposition temperature.

The key factor of CS is to ensure that the particles of feedstock powder impact the substrate at or beyond a critical velocity V_{cr}, which for a given material is the velocity that an individual particle must achieve to successfully deposit onto the substrate due to impact [25].

$$V_{cr} = \sqrt{a\sigma/\rho - bc_p(T_m - T_p)} \tag{4.3}$$

where σ is the temperature dependent flow stress, ρ is the density of the particle, c_p is the heat capacity, T_m is the melting temperature, T_p is the mean temperature of particle upon impact, and a and b are fitting constants. The temperature-dependent flow stress can be approximated as follows

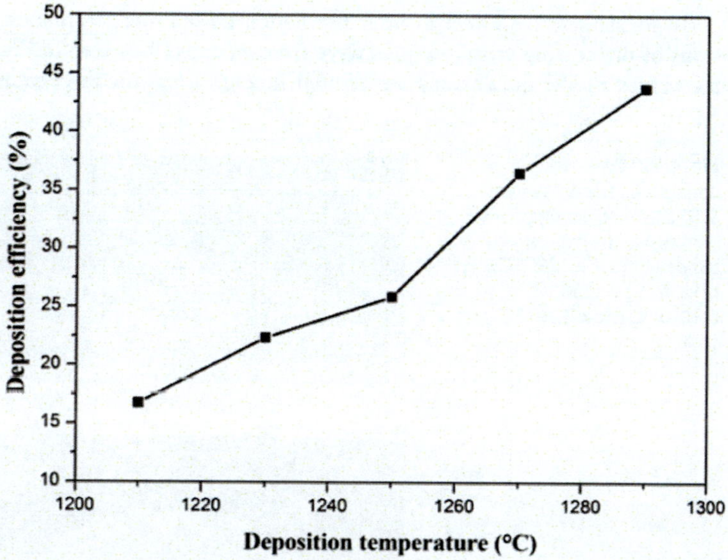

Fig. 4.14 Variation of deposition efficiency with deposition temperature for SLD Stellite 6 coating [10]

$$\sigma = \sigma_{UTS}(1 - \theta) \tag{4.4}$$

$$\theta = (T_p - T_{ref})/(T_m - T_{ref}) \tag{4.5}$$

where θ is the normalized temperature, and T_{ref} is a reference temperature (normally room temperature). For a given material, increasing particle temperature can lead to reduction of the critical velocity according to Eqs. (4.3)–(4.5). In the present CS process experiment, feedstock Stellite 6 powders contain a mixture of particles with various particle sizes and thus a wide spectrum of particle impacting velocities would occur as the spraying particles are accelerated by carrier gas in the convergent-divergent de Laval nozzle. Therefore, only a fraction of particles whose impacting velocities exceed the critical velocity can bond effectively to the substrate. For the arrangement of the laser beam and powder feeding nozzle in Fig. 4.2, the powder jet and the laser beam partially overlapped with each other. Although the spraying particles were travelling at high velocities and had limited time of exposure to laser, it is expected that the particles can be significantly pre-heated in flight by laser prior to impacting the substrate because of high laser energy density and small particle size, which would soften the particles. As a result, the critical velocity could be reduced and the proportion of particles exceeding this velocity would increase, leading to more particles participating in coating formation and thereby improvement in deposition efficiency.

To investigate the effect of laser irradiation on the track formation, the dimensions of the coating cross-sections were measured, as illustrated in Fig. 4.15. Then the values of the dimensions are plotted in Fig. 4.16a for the coatings prepared

Fig. 4.15 Cross-section dimension measurement of SLD Stellite 6 coating prepared at the deposition temperature of **a** 1210 °C, **b** 1230 °C, **c** 1250 °C, **d** 1270 °C and **e** 1290 °C [10]

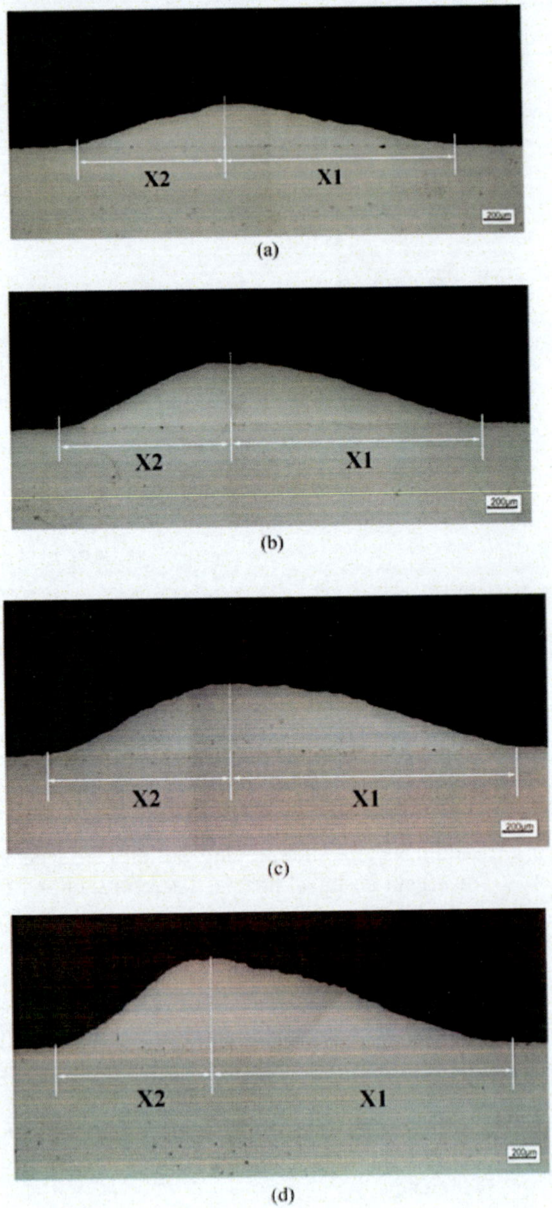

Fig. 4.15 (continued)

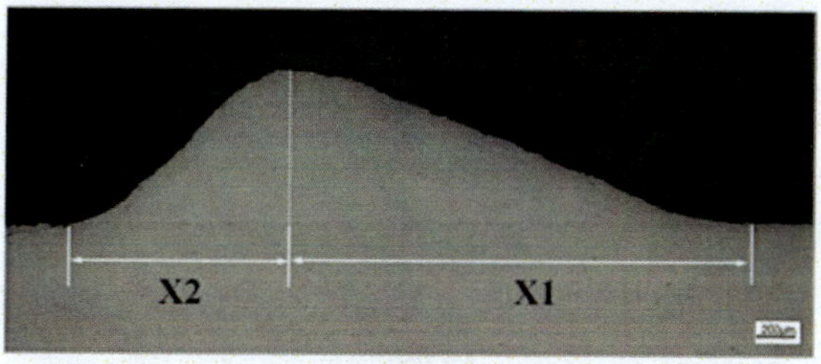

(e)

under different deposition temperatures. It can be found that both track height and width increase with deposition temperature. However, track symmetry decreases with deposition temperature as evidenced by the ratio of X1/X2 shown in Fig. 4.16b. For the SLD Stellite 6 coating, the track shape was determined by not only the particle impacting velocity but also the relative configuration between the powder jet and laser beam. Since the distribution of laser power along the nozzle axis is asymmetric and this would lead to different influences of laser irradiation on the track formation of two sides about the peak of the coating, the asymmetric shape of the coating was caused. This asymmetric shape would have an influence on the overlap of multi-tracks and additive manufacturing process.

4.3.4 Stellite 6 Coating Microstructure

The SEM microstructure of the SLD Stellite 6 coating prepared under the deposition temperature of 1270 °C is shown in Fig. 4.17. At low magnification the interior dendrite structure of feedstock Stellite 6 particles can be clearly observed, but resolidified region around the particles can also be observed, which indicates that the Stellite 6 particles were partially melted at surface by the high temperature in the SLD process and resolidification occurred on the particle surface during cooling. In other words, the Stellite 6 particles were not totally melted. Instead, they had kept interior solidus state during the SLD process. For comparison, the SEM microstructure of the LC Stellite 6 coating is present in Fig. 4.18. Significant difference in microstructure between the two coatings can be found. The Stellite 6 particles were totally melted and formed a dendrite structure in the LC coating. Furthermore, comparing the grain size in the resolidified region (Fig. 4.17b) of the SLD coating with that of the LC coating (Fig. 4.18b), the former is much smaller than the latter.

XRD analysis was performed on the coatings to identify the phases present. As shown in Fig. 4.19, the XRD diffraction patterns indicate that no new phases were generated in the SLD coating. In other words, the original phases (fcc Co solid solution and Cr_7C_3 carbide) of the feedstock powder can be retained from the SLD process. The element distributions in the Stellite 6 coatings were investigated with SEM/EDS. As shown in Fig. 4.20, in the LC coating specimen there is an apparent

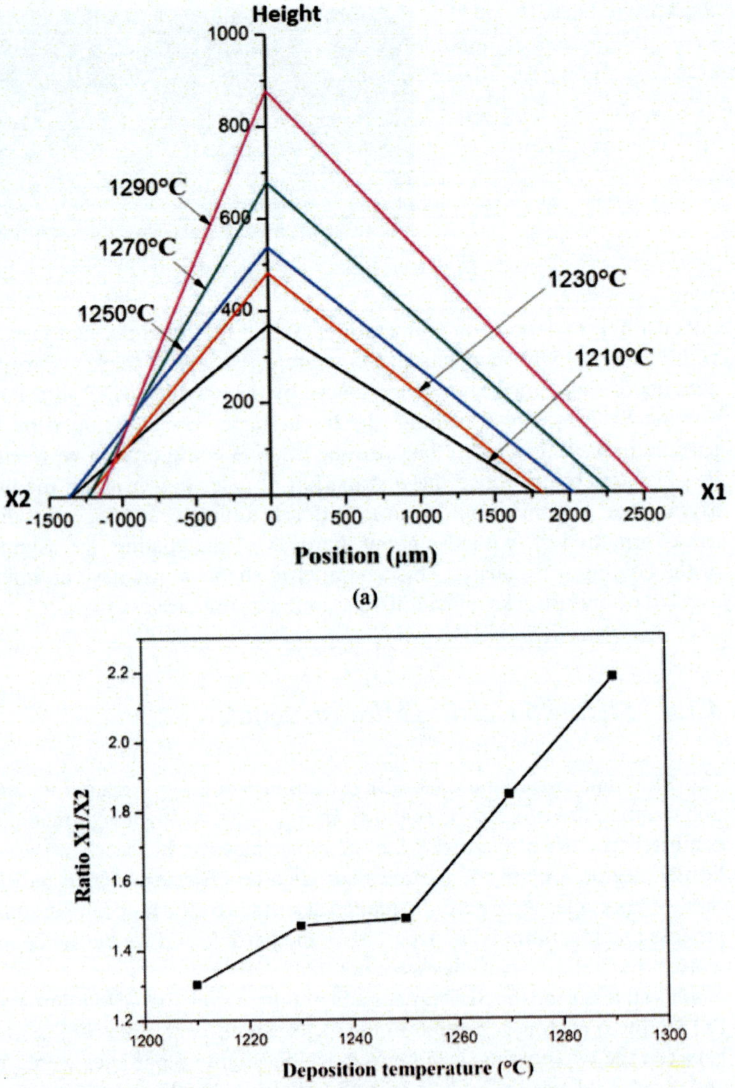

Fig. 4.16 Cross-section dimension measurement results: **a** widths and height and **b** variation of the width ratio with deposition temperature [10]

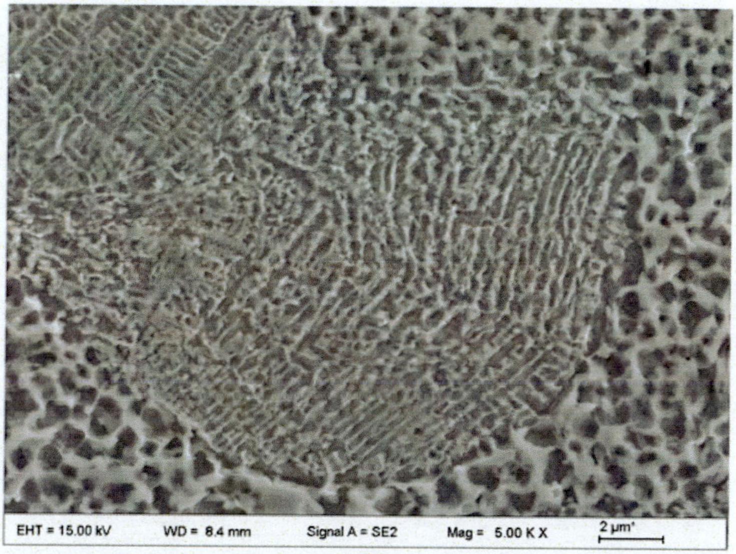

(a)

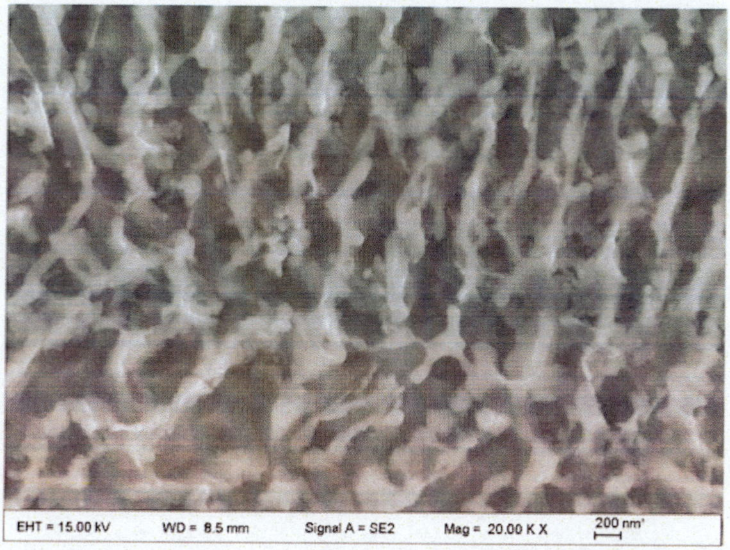

(b)

Fig. 4.17 SEM microstructure of SLD Stellite 6 coating prepared under 1270 °C deposition temperature: **a** at low magnification and **b** at high magnification [10]

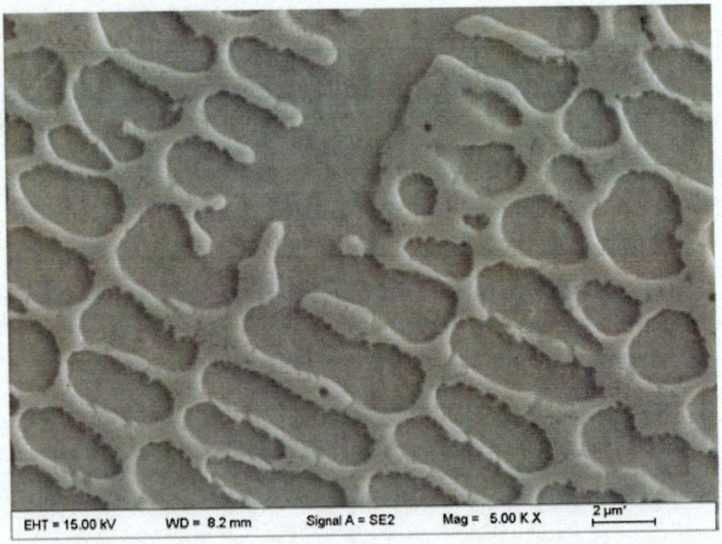

(a)

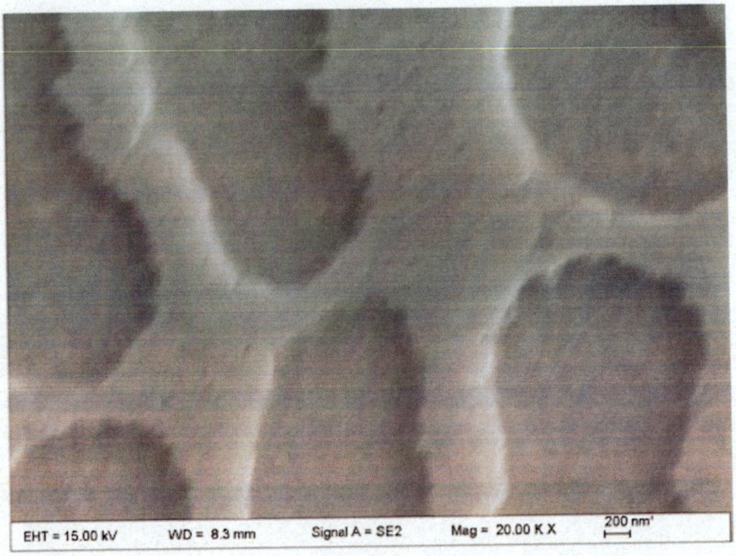

(b)

Fig. 4.18 SEM microstructure of LC Stellite 6 coating: **a** at low magnification and **b** at high magnification [10]

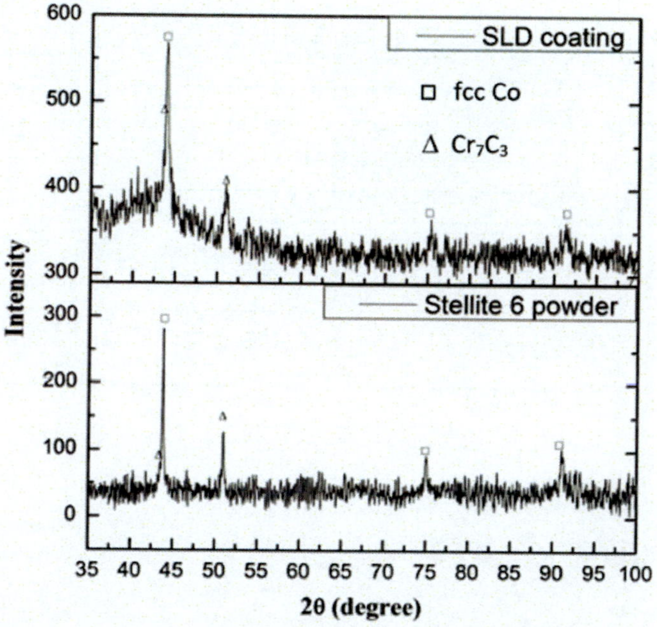

Fig. 4.19 XRD diffraction patterns of Stellite 6 powder and SLD coating [10]

transient region at the interface as highlighted by the rectangle in Fig. 4.20a, where the contents of Co, Cr and Fe elements gradually change from the coating layer to the substrate, whereas there is almost no transient zone in the SLD coating. As highlighted by the rectangle in Fig. 4.20b, the contents of Co, Cr and Fe elements drop down and rise up discontinuously from the coating layer to the substrate. Moreover, it can also be found that the content of Fe element in the LC coating layer is higher than that in the SLD coating layer, which indicates that the elemental dilution in the LC coating is more severe than that in the SLD coating. Dilution of Fe element has a detrimental effect on the hardness, wear, corrosion, and mechanical properties of the coating, which should be minimized.

4.3.5 Stellite 6 Coating Wear Resistance

The pin-on-disk wear test was conducted on the Stellite 6 coatings at room temperature in dry-sliding mode. The pin was a Si_3N_4 ceramic ball having 4 mm diameter and the disk was the coating specimen. The test was performed under a normal load of 500 g at a rotational speed of 500 rpm of the specimen. The evolutions of friction coefficient of the coatings were recorded during the wear tests, thus the variations of friction coefficient with sliding time can be obtained, as plotted in Fig. 4.21. It

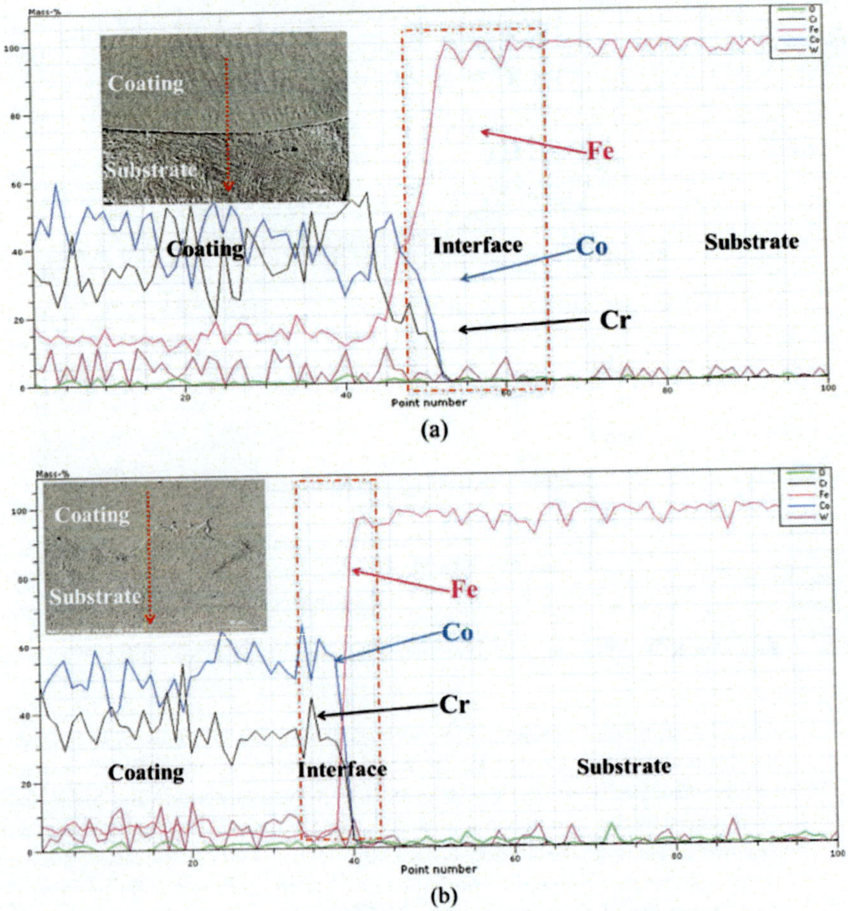

Fig. 4.20 Element distributions in **a** LC specimen and **b** SLD specimen [10]

is shown that the friction coefficient of the SLD coating is smaller and more stable than that of the LC coating. The average friction coefficient of the SLD specimen is about 0.55 after 60 min sliding while that of the LC specimen is about 0.68 after the same sliding time. The worn surfaces of the coating specimens were examined using SEM and the images are given in Fig. 4.22. The surface morphologies of the SLD and LC coatings are similar, both exhibiting abrasive wear characteristics, but the width of the wear track of the SLD coating (590 μm) is smaller than that of the LC coating (688 μm). Also, the LC coating surface appears more damaged with deeper wear scars.

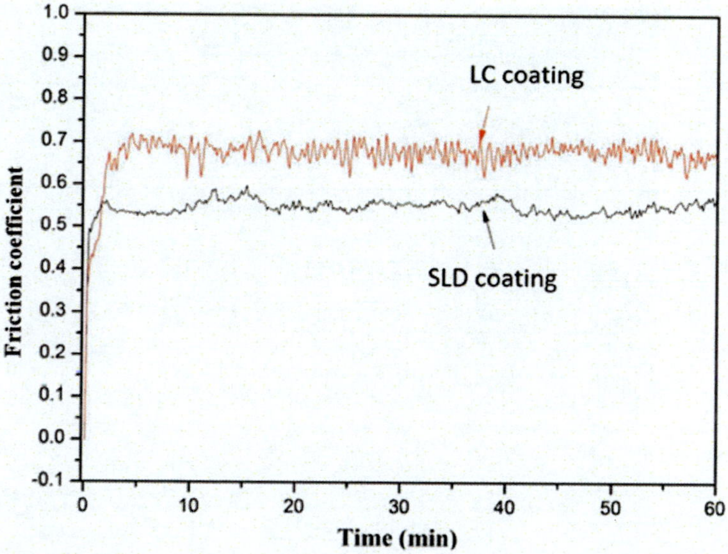

Fig. 4.21 Variation of friction coefficient with sliding time for Stellite 6 coatings [10]

4.3.6 Stellite 6 Coating Corrosion Resistance

Electrochemical corrosion test was conducted on the Stellite 6 coatings using a three-electrode cell electrochemical system (CHI660E) in 1 mol/L H_2SO_4 solution, which was made of 98 wt.% H_2SO_4 with deionized water. The polarization curves were obtained in the tests at a scanning rate of 0.001 V/s, starting from −1.2 V until zero. As shown in Fig. 4.23, for both coatings, the polarization curve has a passivation region, indicating the ability of forming a protective oxide film on the coating surface, but the SLD coating has lager passivation region, implying better corrosion resistance. In addition, the SLD coating has slightly higher corrosion potential (−548 mV) than the LC one (−582 mV), with similar current density, which all indicate that the SLD coating has better corrosion resistance than the LC coating in 1 mol/L H_2SO_4 solution.

4.3.7 Stellite 6 Coating Bonding

Three-point-bending test was performed on the SLD Stellite 6 coating to investigate tensile strength using a computer-controlled tensile testing system (CTM8050, Xieqiang Instrument Co. Ltd.) [11], as schematically shown in Fig. 4.24. The dimensions of the test frame are set as $l = 30$ mm, $L = 20$ mm and $b = d = 1.85$ mm. A point load P was applied normal to the beam top surface at the center of the length

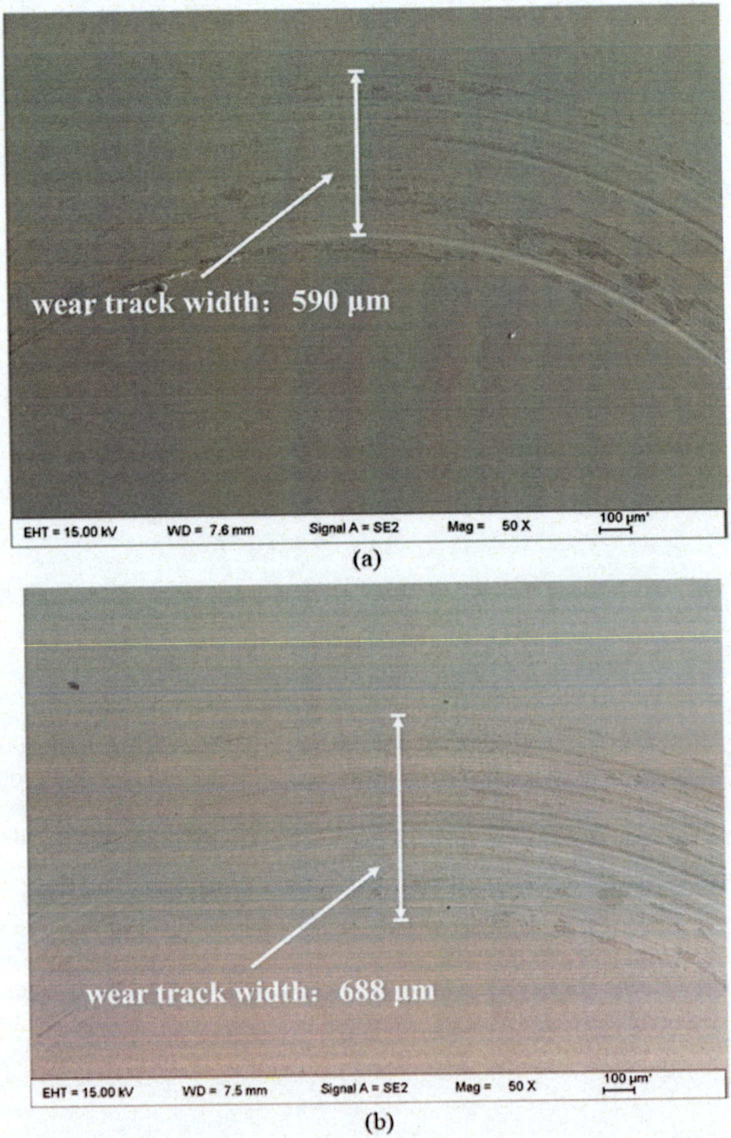

Fig. 4.22 Worn surface of **a** SLD Stellite 6 coating and **b** LC Stellite 6 coating [10]

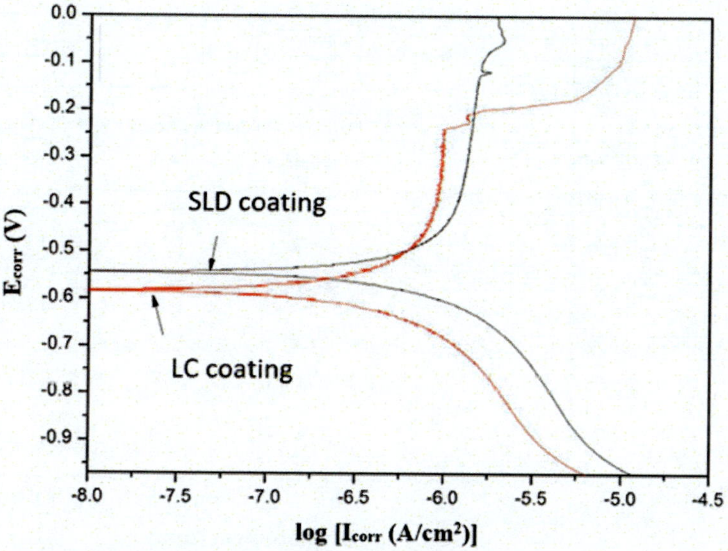

Fig. 4.23 Polarization curves of Stellite 6 coatings tested in 1 mol/L H_2SO_4 solution [10]

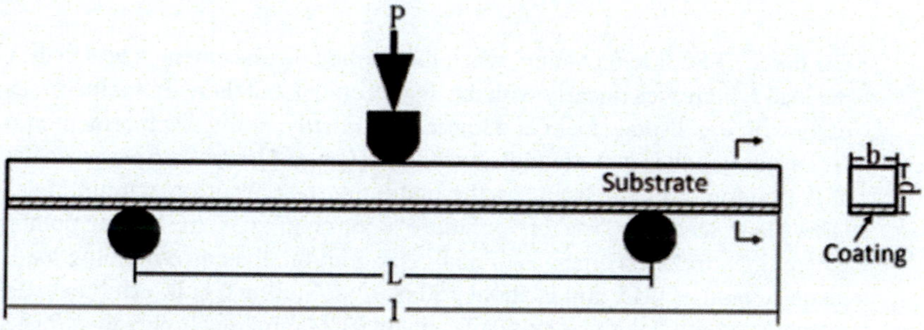

Fig. 4.24 Schematic diagram of three-point-bending test setup [11]

via a hydraulic loading frame at a constant displacement rate of 0.005 mm/s in room temperature environment. The maximum tensile stress is generated at the midpoint of the beam on the outer surface with the coating, which is evaluated by the following formula according to elastic beam bending theory [26].

$$\sigma_f = \frac{3PL}{2bd^2} \tag{4.6}$$

The obtained curve of load P versus applied displacement for the SLD Stellite 6 coating is shown in Fig. 4.25, along the curves of cast Stellite 6 and carbon steel for comparison. It is seen that these curves all have a nearly linear part and nonlinear part.

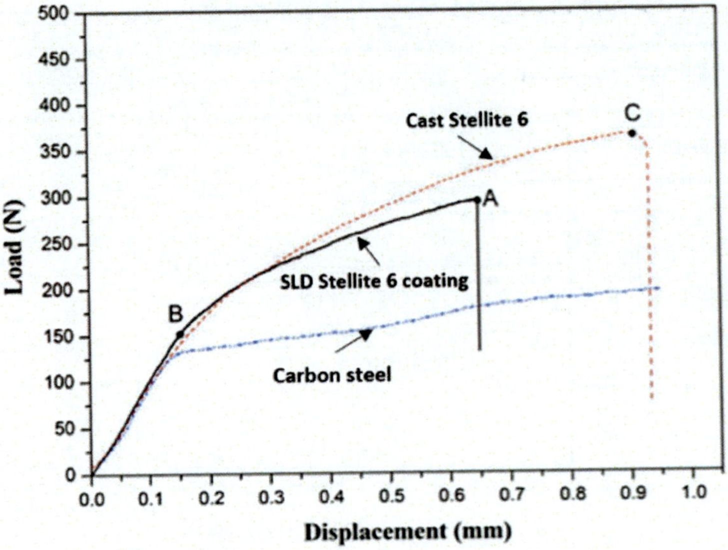

Fig. 4.25 Load *P* versus applied displacement curves of the specimens in three-point-bending test [11]

For the SLD Stellite 6 coating, when the applied displacement is less than 0.15 mm, the load *P* increases linearly with the displacement, but thereafter it increases nonlinearly with the displacement at a lower rate. Finally, at the displacement of 0.65 mm which corresponds to the load *P* about 300 N (point *A*), cracking occurs in the coating at the midpoint of the beam on the bottom surface where maximum tensile stress takes place, leading to fracture or failure of the coating. At the turning point (point *B*) where linear relation changes to nonlinear relation, the corresponding load *P* to the displacement of 0.15 mm is about 150 N. Cast Stellite 6 bulk exhibits better tensile strength (point *C*) with larger displacement (0.94 mm) and higher load *P* (370 N) but carbon steel bulk appears worse in tensile strength, compared with cast Stellite 6 and the SLD Stellite 6 coating. Using Eq. (4.6), the maximum tensile stress generated in the SLD Stellite 6 coating is calculated to be 720 MPa.

The experimental studies have demonstrated that SLD as a novel deposition technique surpasses CS to produce dense coating with good interfacial bonding between particles and between particles and substrate for hard Stellite 6 powder. Deposition temperature has beneficial effect on improving coating surface roughness and deposition efficiency. Because of relatively low laser energy input compared with LC process, the SLD Stellite 6 coating can preserve the original microstructure and phase of the feedstock material, exhibiting better wear/corrosion performance than the LC coating.

4.4 Supersonic Laser Deposition of Nickel Alloy

4.4.1 Ni60 Coating Deposition

Ni60 alloy is one of the most widely used NiCrBSi alloys as self-fluxing powders for conventional laser cladding and thermal spray. Adding Si, B and Cr elements in a nickel-based alloy can effectively reduce its melting point down to 1050 – 1080 °C and enhance the alloy hardness up to HRC62 (734 HV) [27]. The obtained coating exhibits excellent resistance to wear, corrosion and high temperature oxidation over a wide range of temperature [28]. Conventional Ni60 coatings are often prepared via thermal spraying and LC, but Ni60 coatings deposited using these methods exhibit high oxidation and porosity with non-uniform microstructure due to very high temperature during thermal processing [29]. With the viewpoint of temperature, CS technique can solve these problems. However, Ni60 is a very hard material and the coating from CS process has poor adhesion to the substrate. Owing to the unique feature by incorporating laser irradiation in CS process, SLD has been applied to deposit hard Ni60 alloy on carbon steel substrate [14].

Commercially available Ni60 alloy powder with the chemical composition (wt.%) of 18.5% Cr, 14.5% Fe, 0.1% C, 4.5% Si, 3% B, Bal. Ni was used as the coating feedstock. The SEM morphology and size distribution of the Ni60 powder are shown in Fig. 4.26. The Ni60 powder has spherical shape with mean size 18.87 μm. The substrate is medium carbon steel (AISI 1045 steel) with chemical composition (wt.%) of 0.45% C, 0.7% Mn, 0.04% P, 0.05% S, Bal. Fe. The surface of substrate specimen was prepared by blasting using 24 Mesh Al_2O_3, then ultrasonic cleaning in alcohol medium.

The schematic diagram of the SLD system used in this study is given in Fig. 4.2. A high-pressure nitrogen gas (30 bar and 550 °C) supply is split and delivered to a converging nozzle directly and also via a spiral high pressure powder feeder where metal powder particles are entrained. The feedstock stream and carrier gas mix and pass through the nozzle where they are accelerated to supersonic speed. The high-speed flying particles impact a region of the substrate which is synchronously irradiated by the continuous wave diode laser (LDF-4.000, Laserline) with a maximum power of 4 kW. The sprayed zone and laser spot match each other. The SLD zone temperature is monitored by a pyrometer. In this experiment, nitrogen gas was used to accelerate Ni60 powder. The process parameters for the single-track coating and multiple-track coating are as follows: spray distance of 30 mm, traverse rate of 30 mm/s, deposition temperature of 980 °C, laser spot of 5 mm diameter. For comparison, LC was also employed to prepare Ni60 coating on AISI 1045 steel. In the LC process, Ni60 powder was introduced into the beam with a coaxial powder-feed system. The carrier gas of powder-feed and the shielding gas of molten pool were nitrogen and argon, respectively. The laser used in this experiment was the continuous wave diode laser (LDF-2.000, Laserline) with a maximum power of 2 kW.

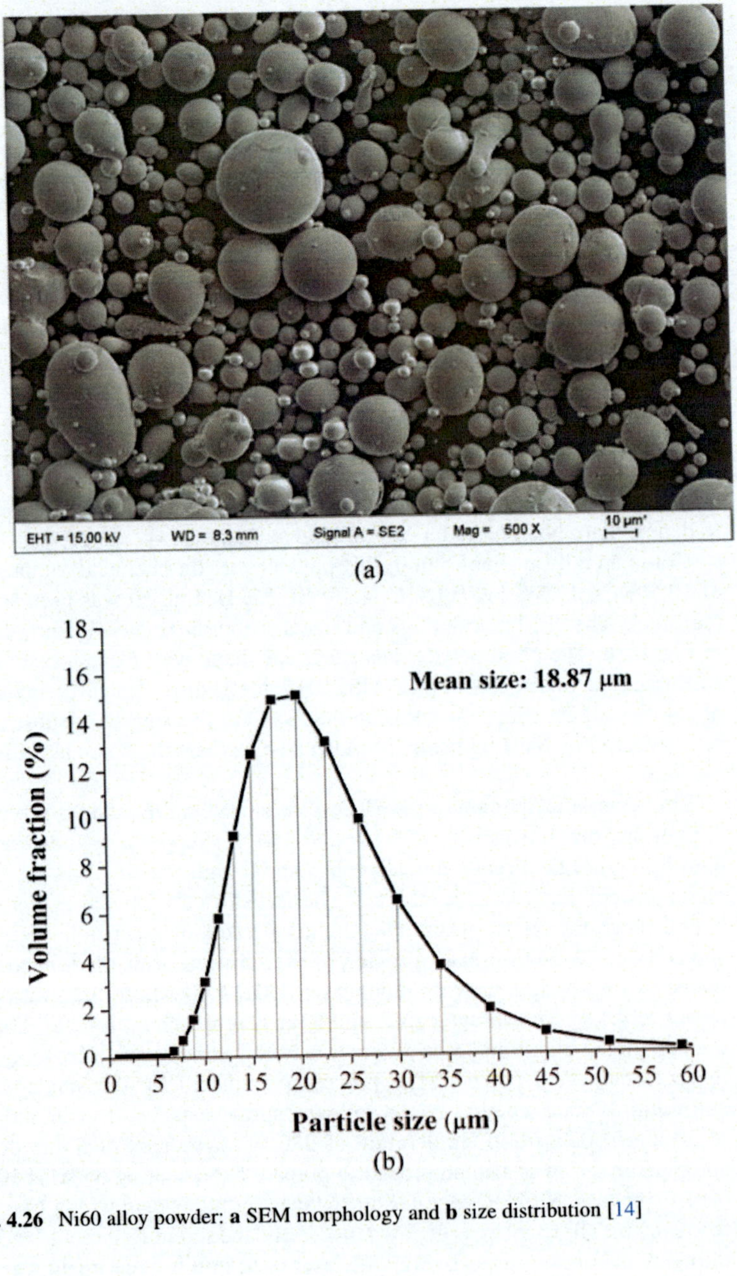

Fig. 4.26 Ni60 alloy powder: **a** SEM morphology and **b** size distribution [14]

The process parameters for preparing Ni60 coating were feeding rate of 15 g/min, scanning velocity of 10 mm/s, laser power of 1.3 kW, and laser spot of 4 mm diameter.

4.4.2 Ni60 Coating Microstructure

The microstructures of the SLD and LC Ni60 coatings were examined with SEM/EDS. As shown in Figs. 4.27a and 4.28a, both SLD and LC Ni60 coatings have three distinct zones in cross-section: coating layer, heat-affected-zone (HAZ) and substrate, but the interfaces between these zones are flat for the SLD coating, while those for the LC coating are curved. Regarding the microstructures of the coating layers, the SLD coating has the characteristics of solid-state deposition structure, that is, the feedstock Ni60 particles are observed, as shown in Fig. 4.27b. On the contrary, the LC coating shows typical dendrite structure (Fig. 4.28b). At high magnification, the microstructures of both coatings have two distinct areas, as marked in Figs. 4.27c and 4.28c. EDS analysis was performed on these areas and the results are reported in Table 4.2. It is shown that areas 1 and 3 contain higher Cr and Si contents and lower Ni content. Based on the observation of previous research [30], these areas contain Cr-rich carbides/borides and Ni-rich silicide. Areas 2 and 4 contain high Ni so that they are γ-Ni solid solution. These results indicate that the Ni60 coatings have similar microstructure with respect to phases formed, but they have different morphologies, as seen in the SEM images (Figs. 4.27b and 4.28b). In the SLD process, feedstock Ni60 particles were not melted or may only be partially melted at surface, and the coating was formed mainly due to the plastic deformation of the particles and substrate. As a result, the coating microstructure exhibits the feature of bonded solid particles with the feedstock Ni60 alloy microstructure unchanged. However, for the LC coating, since all Ni60 particles were totally melted in the LC process and the Ni60 molten pool experienced resolidification during cooling, forming a dendritic structure. In addition, the high laser heat also melted the substrate surface, causing dilution and formation of the curved interfaces, as seen in Fig. 4.28a. Furthermore, it is noted that the SLD coating has finer microstructure than the LC one. This is due to the fact that the SLD coating kept the original microstructure of feedstock Ni60 powder, which has finer grains ascribing to high cooling rate during the gas atomization process of Ni60 powder.

XRD analysis was conducted on feedstock Ni60 powder and the coating specimens to further identify the phases in the microstructures. The analysis experiments were operated at 40 kV and 40 mA to generate monochromatic Cu K_α radiation with a wavelength of 0.154 nm. Form the XRD pattern Fig. 4.29, γ-Ni, $FeNi_n$, Cr_mC_n, Cr_3B and Ni_3Si_2 are the main phases in the SLD coating, and the Cr_mC_n and Cr_3B are the main strengthening phases of feedstock Ni60 powder. However, the phases in the XRD spectrum of the LC coating differ from that of the SLD coating in that the LC process generated a new phase Fe_5C_2 due to dilution where excessive C and Fe contents diffused into the Ni60 coating from the substrate. Based on the EDS and XRD results in Table 4.2 and Fig. 4.29, in the SLD coating area 1 (Fig. 4.27c) mainly

(a)

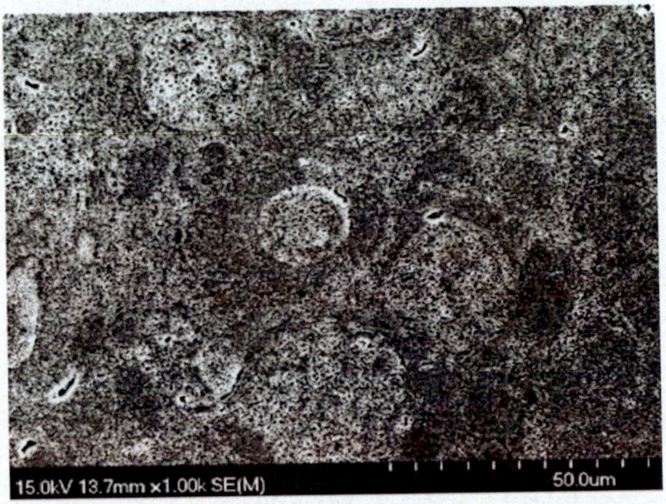

(b)

Fig. 4.27 SEM microstructure of SLD Ni60 coating: **a** cross-section, **b** at low magnification and **c** at high magnification [14]

consist of Cr_mC_n, Cr_3B, Ni_3Si_2, area 2 (Fig. 4.27c) is γ-Ni solid solution; while in the LC coating area 3 (Fig. 4.28c) is composed of Cr_5Si_3, Fe_5C_2, Cr_mC_n, the area 4 (Fig. 4.28c) is γ-Ni solid solution. Figure 4.29 also shows that the phases in the SLD coating are similar to that of feedstock Ni60 powder. This confirmed that the SLD Ni60 coating has retained the original microstructure of feedstock Ni60 powder and exhibits solid-state deposition characteristics.

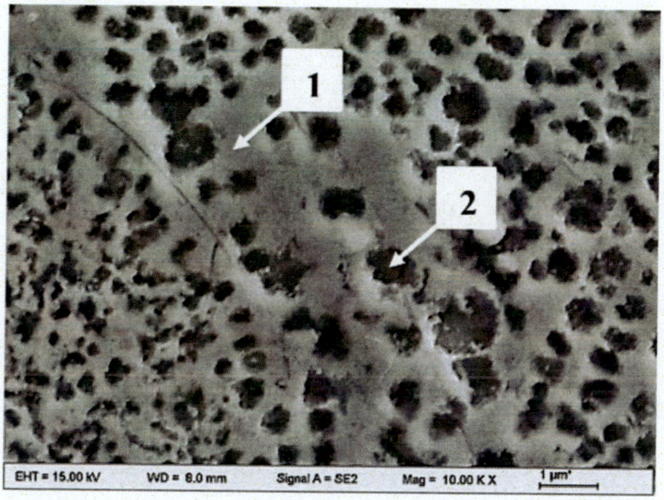

(c)

Fig. 4.27 (continued)

4.4.3 Ni60 Coating Dilution

The microstructural analyses show the difference between the SLD and LC coating, which may be caused by dilution of the substrate material. The line scanning of Ni, Cr, Si and Fe elements from the coating to the substrate was performed on the cross-sections of the coating specimens with EDS. As shown in Fig. 4.30, Ni, Cr and Si contents in the SLD coating are obviously higher than that in the LC one, while the Fe content of the LC coating is much higher than that of the SLD one, indicating that Fe dilution occurred more in the former than in the latter. The noticeable difference in dilution behavior between the two coatings can be found at the coating/substrate interfaces, as shown in Fig. 4.30. For the SLD coating specimen there is no transient zone of element distributions, that is, the contents of Ni, Cr, Si and Fe elements drop down and rise up discontinuously from the coating to the substrate, whereas there is an obvious transient zone at the interface where the contents of Ni, Cr, Si and Fe elements gradually change from the coating to the substrate of the LC coating specimen. This was caused by elemental diffusion which was promoted by the convection and stirring of metal elements in the molten pool, as reported by other researchers [31].

(a)

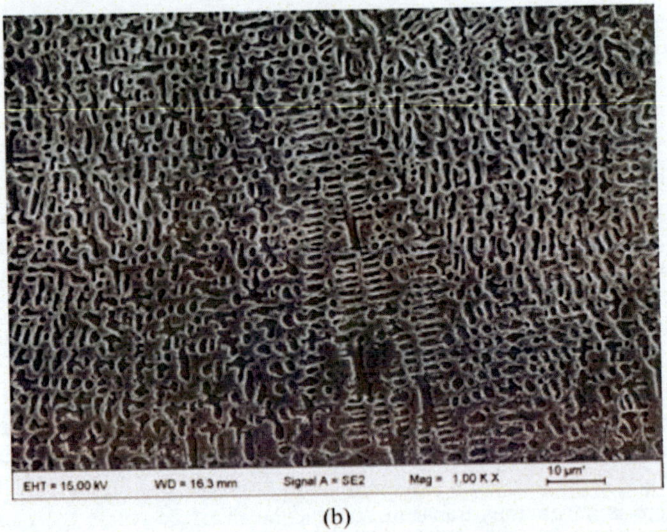

(b)

Fig. 4.28 SEM microstructure of LC Ni60 coating: **a** cross-section, **b** at low magnification and **c** at high magnification [14]

4.4.4 Ni60 Coating Hardness

The hardness of the coatings was evaluated from the coating surface to the substrate at a constant interval of 0.06 mm using a Digital Micro-Hardmeter (HMV-2 T, SHIMADZU), under an indentation load of 3 N ($HV_{0.3}$). As shown in Fig. 4.31,

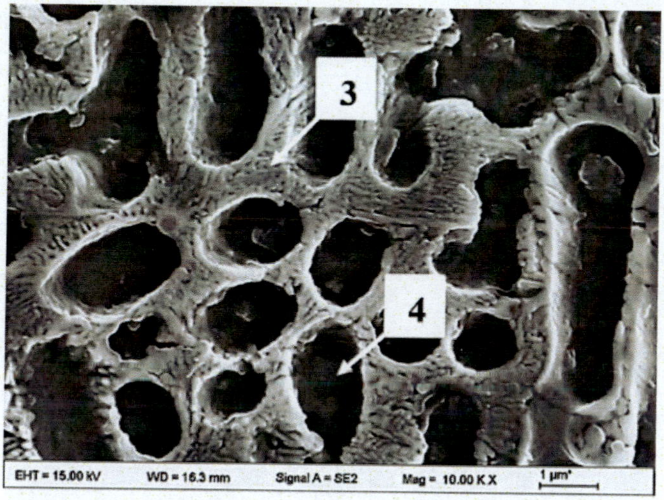

(c)

Fig. 4.28 (continued)

Table 4.2 Elemental concentrations (wt.%) from the EDS analyses of the selected areas in the microstructures of Ni60 coatings [14]

Area	Si	Cr	Fe	Ni
1	6.15	23.10	12.11	54.31
2	4.56	14.51	19.31	60.43
3	2.83	11.34	53.05	15.77
4	1.04	8.87	47.56	27.83

the average hardness of the SLD coating is 867 ± 24 $HV_{0.3}$ and that of the LC one is 625 ± 55 $HV_{0.3}$. The hardness of the SLD coating appears less varying through the coating layer than the LC one, but it decreases remarkably across the coating/substrate interface within the HAZ of the substrate. As for the LC coating hardness shows a moderate decreasing trend, which is defined as gradient character. For both coatings hardness decreases gradually from bonding interface to the substrate, reaching the hardness value of the substrate material (steel), but the LC coating has larger HAZ than the SLD one. The higher hardness of the SLD coating is attributed to more hard CrB ceramic and Cr_mC_n carbides phases precipitated within the γ-Ni solid solution [32].

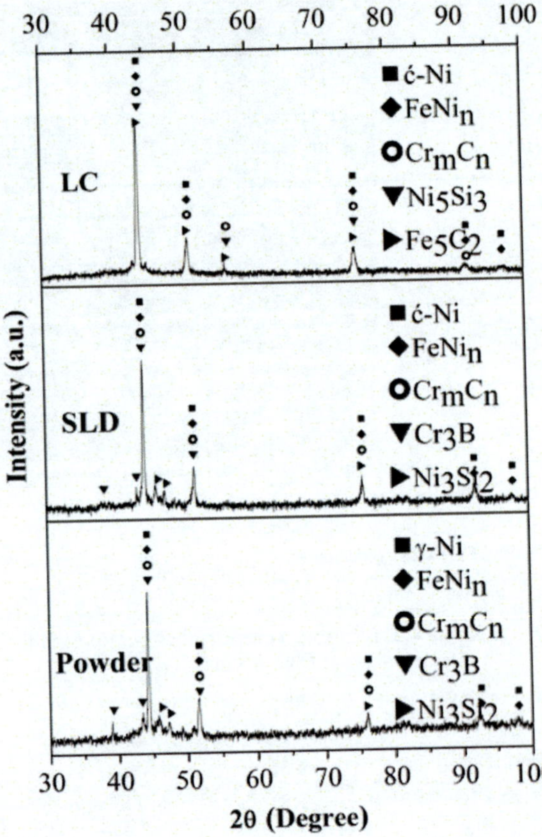

Fig. 4.29 XRD diffraction patterns of feedstock Ni60 powder and Ni60 coatings [14]

4.4.5 Ni60 Coating Wear Resistance

The pin-on-disc wear test was conducted on the SLD and LC Ni60 coatings at room temperature in dry sliding condition. The pin was a Si_3N_4 ceramic ball with 4 mm diameter and the disk was the coating specimen. The test was performed under a normal load of 500 g at a rotational speed of 500 rpm of the disk between the contacting surfaces. The test duration was 60 min for each specimen and the diameter of the generated wear track was 5 mm. The wear loss was evaluated using a three-dimensional microscopic system super depth of field (VHX-5000) by simulating the cross-section profile of the wear track (Fig. 4.32). The volume loss was estimated by calculating the volume of the wear track, which is the area of the cross-sectional area multiplied by the periphery of the wear track. The wear loss results of the two coatings are reported in Table 4.3. Evidently, the SLD coating has better wear resistance than the LC one.

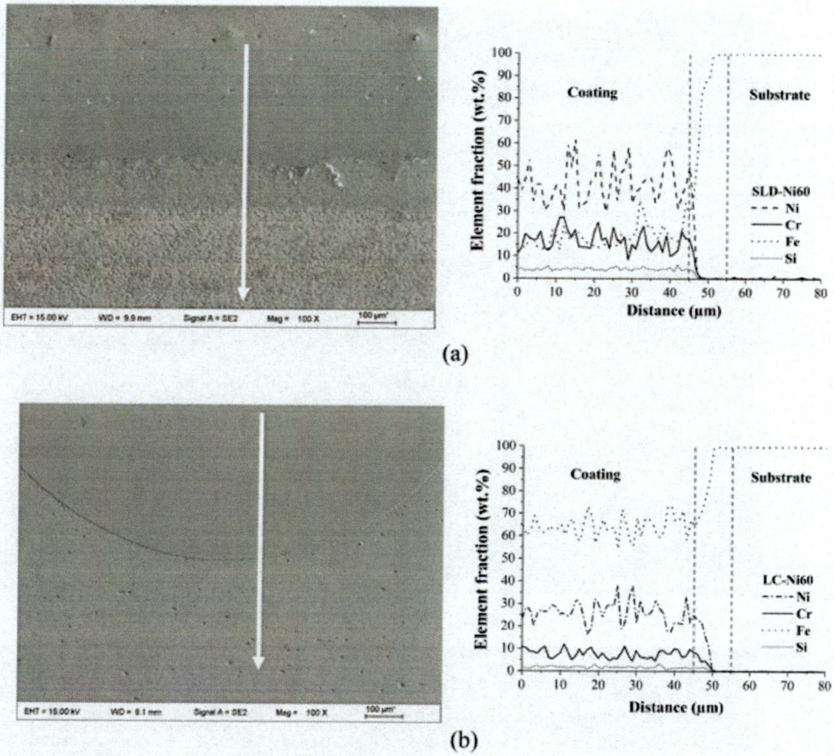

Fig. 4.30 Dilution analysis of **a** SLD Ni60 coating and **b** LC Ni60 coating [14]

Fig. 4.31 Hardness profiles of Ni60 coatings [14]

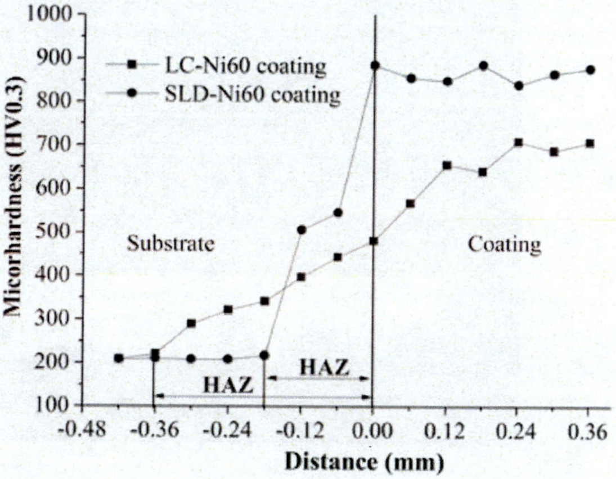

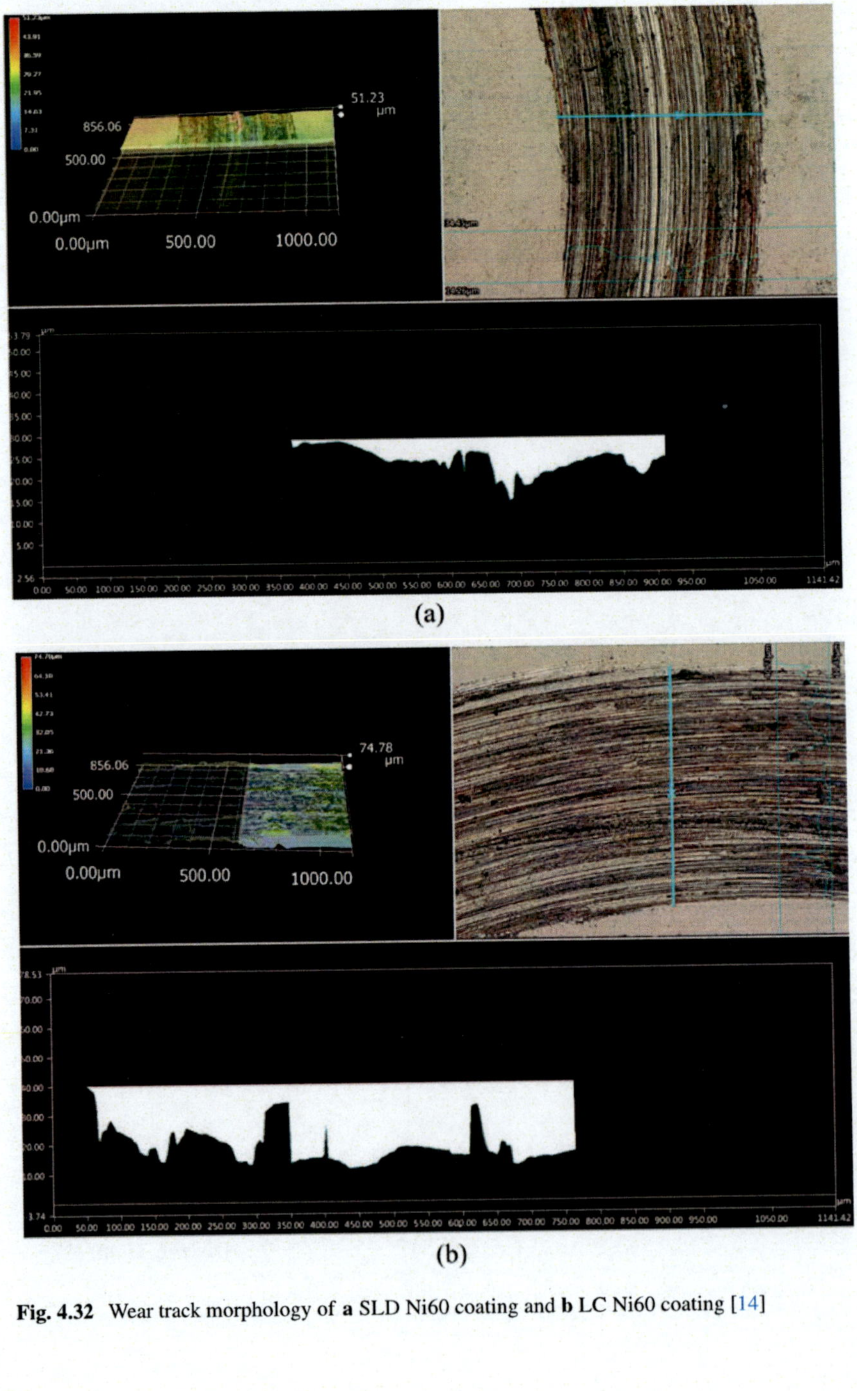

Fig. 4.32 Wear track morphology of **a** SLD Ni60 coating and **b** LC Ni60 coating [14]

Table 4.3 Pin-on-disk wear test results of Ni60 coatings [14]

	Volume loss (mm^3)	Maximum depth of wear track (μm)
SLD coating	0.46 ± 0.06	21.3 ± 3.1
LC coating	0.98 ± 0.08	30.6 ± 2.8

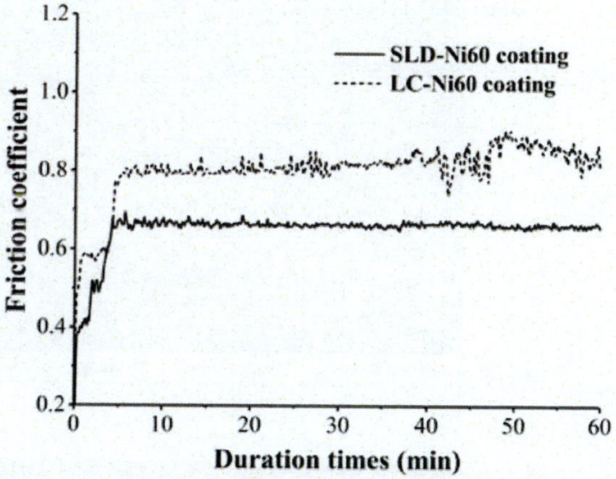

Fig. 4.33 Variations of friction coefficient with sliding time of Ni60 coatings [14]

The friction evolutions of the coating specimens were recorded during the wear tests, then the friction coefficients versus time can be obtained, as plotted in Fig. 4.33. The friction coefficient of the SLD coating is much smaller and more stable than that of the LC coating. The average coefficient of the SLD coating is about 0.68 after 60 min sliding, but the average coefficient of the LC coating is about 0.82 at the same time. The wear loss and friction coefficient results agree well.

To further explore the wear mechanisms of the Ni60 coatings, the worn surfaces of the coating specimens were examined using SEM, with the images shown in Fig. 4.34. The wear track of the LC coating shows deeper scars or plough than that of the SLD coating and the width of the wear track of the LC coating is wider than that of the SLD one. On both surfaces, there are two distinct areas in dark and light, respectively, as indicated in Fig. 4.34; the dark area is more in the LC coating than in the SLD coating. EDS analysis was performed on the dark and light areas of the wear tracks, with the results presented in Fig. 4.35. It is shown that the dark areas (2 and 4) contain Ni, Cr, Fe, Si, C and high content of O, but O is hardly detected in the light areas (1 and 3). This implies that oxidation had occurred in the dark area of the worn track when the coating specimen was under the sliding wear tests. More dark areas in the worn surface of the SLD coating indicate more friction heat generated in the surface during wear, which is consistent with the friction observations in Fig. 4.33.

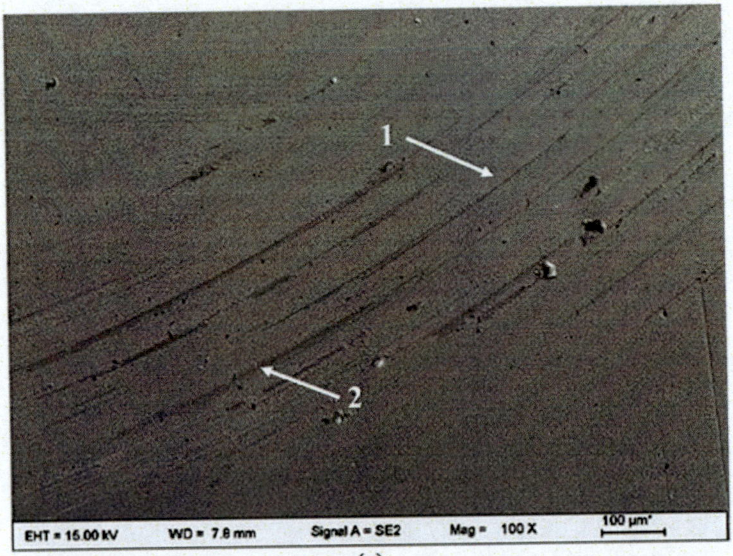

(a)

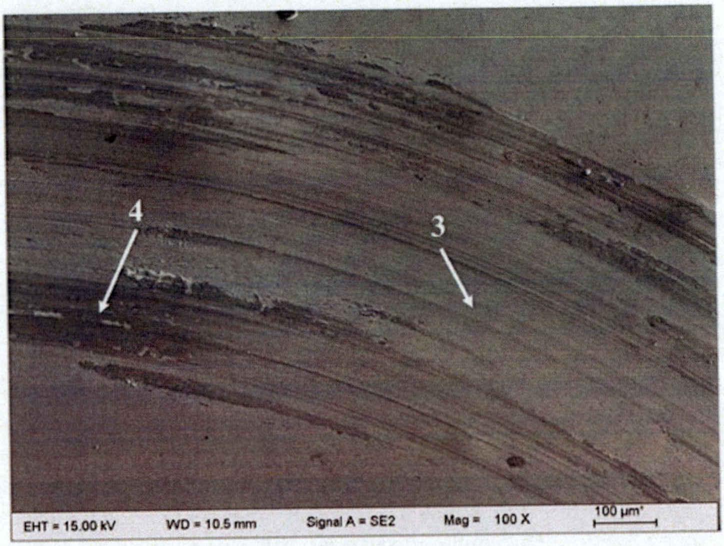

(b)

Fig. 4.34 SEM worn surface morphology of **a** SLD Ni60 coating and **b** LC Ni60 coating [14]

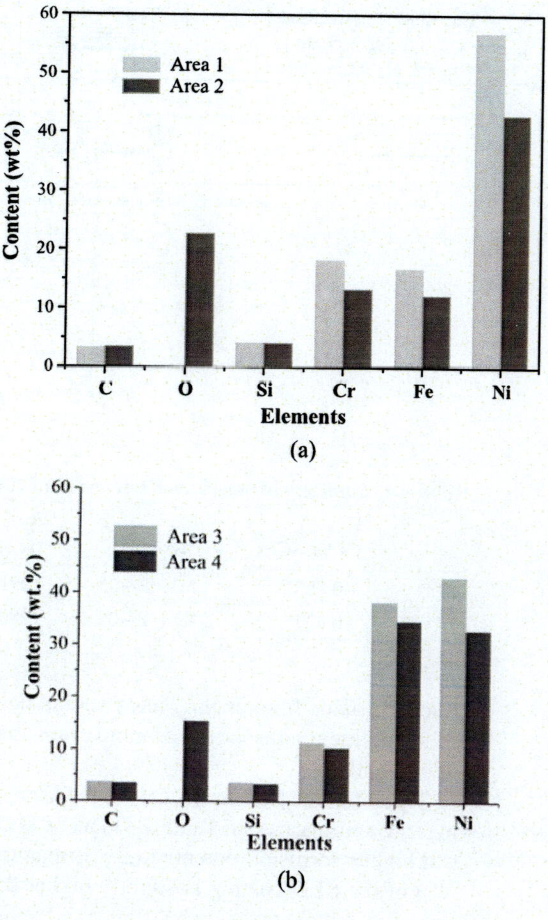

Fig. 4.35 Elemental concentrations of EDS analysis for worn surface of **a** SLD Ni60 coating and **b** LC Ni60 coating [14]

4.4.6 Ni60 Coating Corrosion Resistance

Electrochemical corrosion test was conducted on the Ni60 coatings using an electrochemical system (CHI660E) in 1 mol/L H_2SO_4 solution. The experimental procedures and equipment were the same as used for the SLD Stellite 6 coating, as detailed previously. The polarization curves for the Ni60 coatings are presented in Fig. 4.36. It can be seen that the LC coating has an obvious passivation region, indicating a protective oxide film formed on the specimen surface, due to presence of high Cr content in Ni60 alloy, but the SLD coating only exhibits the tendency of passivation. This implies that the former has better behavior of oxide film than the latter. As discussed earlier, the SLD coating has finer grains than the LC one in microstructure,

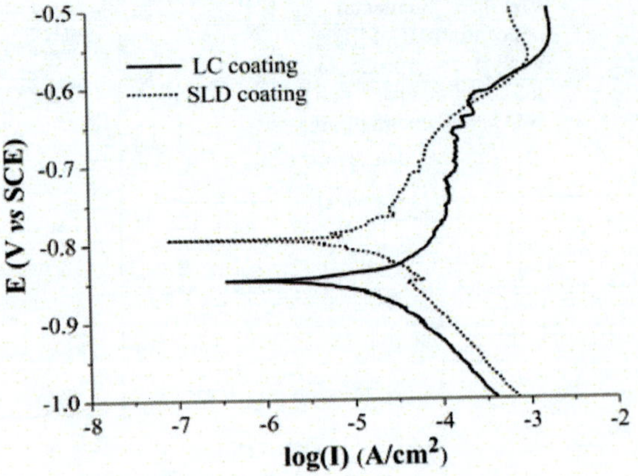

Fig. 4.36 Polarization curves of Ni60 coatings in 1 mol/L H_2SO_4 solution [14]

Table 4.4 Summary of polarization test results of Ni60 coatings in 1 mol/L H_2SO_4 solution [14]

	E_{corr} (V vs. SCE)	I_{corr} ($\mu A/cm^2$)	R_p ($\Omega.cm^2$)	β_a (V/decade)	β_c (V/decade)
SLD	−0.795	2.046E−05	1091	13.94	6.67
LC	−0.847	3.147E−05	1080	5.34	7.68

since the oxide film on Ni60 alloy should be a result of Cr oxidation from the solid solution, finer grains introduce more grain boundaries, thus resulting in smaller areas of oxide film.

The corrosion parameters calculated from the polarization curves of the Ni60 coatings are summarized in Table 4.4. The SLD coating has lower current density (I_{corr}) and higher corrosion potential (E_{corr}) than the LC one. The polarization resistance (R_p) of the SLD coating is slightly higher than that of the LC one. These parameters show better corrosion resistance of the SLD coating, that is, better resistance to electron transferring in the electrochemical reaction.

Hard Ni60 alloy powder can be deposited on medium carbon steel forming a dense and well-bonded coating via SLD process or LC process. Since the energy input rate of SLD process is lower than that of LC process, the SLD Ni60 coating exhibits special characteristics, such as unchanged fine microstructure and phases of feedstock Ni60 powder, and low dilution rate. With these features the SLD Ni60 coating shows superior performance of microhardness, wear resistance and corrosion resistance in 1 mol/L H_2SO_4 solution, to the LC coating.

4.5 Summary of Supersonic Laser Deposition of Metals

Cold spray (CS) is an emerging material deposition technique by which metallic particles are deposited onto substrate to form a coating at a temperature well below the melting point of sprayed materials. During the CS process, the deposited particles are accelerated to a high velocity in a supersonic gas flow and then impinge on the substrate to form the coating. The coating microstructure and properties are greatly influenced by particle preheating and substrate softening.

Laser, as a high-heat source, has been widely used in various surface enhancements such as laser cladding (LC), laser hardening, etc. It can also be combined with other techniques, for example, cold spray, to produce high performance coatings. In other words, CS process can be improved with assistance of synchronous laser irradiation. A novel deposition technique, known as supersonic laser deposition (SLD), has been developed by introducing laser into CS process. It was applied to depositing copper on stainless steel substrate. The influence of synchronous laser irradiation on the Cu coating characteristics was investigated. The results showed that the coating surface with laser irradiation was smoother than that without laser irradiation. The peak coating thickness increased by about 70% as synchronous laser irradiation was employed, indicating an improvement in deposition efficiency. It was also found that with synchronous laser irradiation the coating was denser and the coating/substrate interfacial bonding was better as compared with that without laser irradiation. Moreover, the EDS and XRD analyses found Cu oxidation occurring in the SLD coating, but the oxide was trivial. The aforementioned improvements on the coating largely arose from particle preheating and substrate softening by synchronous laser irradiation in the CS process.

Depositing hard materials such as Stellite 6 solely by CS is a challenge due to limited ability in plastic deformation of the materials, but it can be facilitated by using SLD. The experimental results demonstrated that the surface undulation was reduced and deposition efficiency increased with laser irradiation temperature due to the softening effect on the Stellite 6 particles and substrate generated by laser heating. The as-deposited Stellite 6 tracks showed asymmetric shapes which were influenced by the relative configuration of powder stream and laser beam. The SLD coatings could preserve the original microstructure and phase of the feedstock material due to relatively low laser energy input, which resulted in the superior wear/corrosion-resistant properties as compared with the counterpart prepared by laser cladding (LC).

Hard Ni60 alloy powder was deposited successfully on medium carbon steel substrate via SLD process. Since the energy inputting rate of the SLD process is lower than that of the LC process, the Ni60 coating with SLD exhibited special characteristics, such as unchanged fine powder microstructure, unchanged phases and low dilution rate. With these features the SLD Ni60 coating displayed superior performance of microhardness and wear resistance to the LC one. In the 1 mol/L H_2SO_4 solution, the SLD coating showed better corrosion resistance than the LC one. The SLD process also suppressed the dilution of the steel substrate.

References

1. W. Wong, P. Vo, E. Irissou, A.N. Ryabinin, J.G. Legoux, S. Yue, Effect of particle morphology and size distribution on cold-sprayed pure titanium coatings. J. Therm. Spray Technol. **22**(7), 1140–1153 (2013)
2. T. Suhonen, T. Varis, S. Dosta, M. Torrell, J.M. Guilemany, Residual stress development in cold sprayed Al Cu and Ti Coatings. Acta Mater. **61**, 6329–6337 (2013)
3. B. AL-Mangour, R. Mongrain, E. Irissou, S. Yue, Improving the strength and corrosion resistance of 316L stainless steel for biomedical applications using cold spray. Surf. Coat. Technol. **216**, 297–307 (2013).
4. Q. Zhang, C.J. Li, X.R. Wang, Z.L. Ren, C.X. Li, G.J. Yang, Formation of NiAl intermetallic compound by cold spraying of ball-milled Ni/Al alloy powder through postannealing treatment. J. Therm. Spray Technol. **17**(5–6), 715–720 (2008)
5. A.S. Alhulaifi, G.A. Buck, W.J. Arbegast, Numerical and experimental investigation of cold spray gas dynamic effects for polymer coating. J. Therm. Spray Technol. **21**(5), 852–862 (2012)
6. S. Kikuchi, S. Yoshino, M. Yamada, M. Fukumoto, K. Okamoto, Microstructures and thermal properties of cold-sprayed Cu-Cr composite coatings. J. Therm. Spray Technol. **22**(6), 926–931 (2013)
7. X.T. Luo, G.J. Yang, C.J. Li, K. Kondoh, High strain rate induced localized amorphization in cubic BN/NiCrAl nanocomposite through high velocity impact. Scripta Mater. **65**, 581–584 (2011)
8. M. Bray, A. Cockburn, W. O'Neill, The laser-assisted cold spray process and deposit characterisation. Surf. Coat. Technol. **203**, 2851–2857 (2009)
9. B. Li, L.J. Yang, Z.H. Li, J.H. Yao, Q.L. Zhang, Z.J. Chen, G. Dong, L. Wang, Beneficial effects of synchronous laser irradiation on the characteristics of cold-sprayed copper coatings. J. Therm. Spray Technol. **24**(5), 836–847 (2015)
10. B. Li, Y. Lin, J.H. Yao, Z.H. Li, Q.L. Zhang, X. Zhang, Influence of laser irradiation on deposition characteristics of cold sprayed Stellite 6 coatings. Opt. Laser Technol. **100**, 27–39 (2018)
11. J.H. Yao, Z.H. Li, B. Li, L.J. Yang, Characteristics and bonding behavior of Stellite 6 alloy coating processed with supersonic laser deposition. J. Alloy. Comp. **661**, 526–534 (2016)
12. F. Luo, A. Cockburn, R. Lupoi, M. Sparks, W. O'Neill, Performance comparison of Stellite 6 deposited on steel using supersonic laser deposition and laser cladding. Surf. Coat. Technol. **212**, 119–127 (2012)
13. F. Luo, A. Cockburn, D.B. Cai, M. Sparks, Y.H. Lu, C.R. Ding, R. Langford, W. O'Neill, J.H. Yao, R. Liu, Simulation analysis of Stellite 6 particle impact on steel substrate in supersonic laser deposition process. J. Therm. Spray Technol. **24**(3), 378–393 (2015)
14. J.H. Yao, L.J. Yang, B. Li, Z.H. Li, Characteristics and performance of hard Ni60 alloy coating produced with supersonic laser deposition technique. Mater. Des. **83**, 26–35 (2015)
15. L.J. Yang, B. Li, J.H. Yao, Z.H. Li, Effects of diamond size on the deposition characteristic and tribological behavior of diamond/Ni60 composite coating prepared by supersonic laser deposition. Diamond Related Mater. **58**, 139–148 (2015)
16. J.H. Yao, L.J. Yang, B. Li, Z.H. Li, Beneficial effects of laser irradiation on the deposition process of diamond/Ni60 composite coating with cold spray. Appl. Surf. Sci. **330**, 300–308 (2015)
17. B. Li, J.H. Yao, Q.L. Zhang, Z.H. Li, L.J. Yang, Microstructure and tribological performance of tungsten carbide reinforced stainless steel composite coatings by supersonic laser deposition. Surf. Coat. Technol. **275**, 58–68 (2015)
18. B. Li, Y. Lin, J.H. Yao, Z.H. Li, Q.L. Zhang, Solid-state fabrication of WC_P-reinforced Stellite 6 composite coatings with supersonic laser deposition. Surf. Coat. Technol. **321**, 386–396 (2017)
19. A. Papyrin, V. Kosarev, K.V. Klinkov, V.M. Fomin, *Cold Spray Equipments and Technologies, Cold Spray Technology* (Elsevier Ltd., Oxford, 2007), pp. 179–247
20. H. Assadi, F. Gartner, T. Stoltenhoff, H. Kreye, Bonding mechanism in cold gas spraying. Acta Mater. **51**, 4379–4394 (2003)

21. C.J. Li, W.Y. Li, H.L. Liao, Examination of the critical velocity fro deposition of particles in cold spraying. J. Thermal. Spray Technol. **15**, 212–222 (2006)
22. C.J. Li, W.Y. Li, Y.Y. Wang, Formation of metastable phases in cold-sprayed soft metallic deposit. Surf. Coat. Technol. **198**, 469–473 (2005)
23. E. Calla, D.G. McCartney, P.H. Shipway, Effect of deposition conditions on the properties and annealing behavior of cold-sprayed copper. J. Thermal. Spray Technol. **15**(2), 255–262 (2006)
24. J.R. Davis, *Cobalt-Base Alloys, in Nickel, Cobalt, and Their Alloys* (Materials Park, ASM International, 2000), pp. 362–406
25. T. Schmit, F. Gärtner, H. Assadi, H. Kreye, Development of a generalized parameter window for cold spray deposition. Acta Mater. **54**, 729–742 (2006)
26. A.N. Adsul, S.G. Chavan, P.N. Gore, Review of theories regarding material bending. Inter. J. Sci. Eng. Res. **4**(10), 1293–1299 (2013)
27. X.C. Zhang, B.S. Xu, S.T. Tu, F.Z. Xuan, H.D. Wang, Y.X. Wu, Effect of spraying power on the microstructure and mechanical properties of supersonic plasma-sprayed Ni-based alloy coatings. Appl. Surf. Sci. **254**, 6318–6326 (2008)
28. S. Hong, Y.P. Wu, G.Y. Li, B. Wang, W.W. Gao, G.B. Ying, Microstructural characteristics of high-velocity oxygen-fuel (HVOF) sprayed nickel-based alloy coating. J. Alloy. Compd. **581**, 398–403 (2013)
29. H.J. Kim, S.Y. Hwang, C.H. Lee, P. Juvanon, Assessment of wear performance of flame sprayed and fused Ni-based coatings. Surf. Coat. Technol. **172**, 262–269 (2003)
30. C. Navas, R. Vijande, J.M. Cuetos, M.R. Fernandez, J. de Damborenea, Corrosion behaviour of NiCrBSi plasma-sprayed coatings partially melted with laser. Surf. Coat. Technol. **201**, 776–785 (2006)
31. F. Luo, A. Cockburn, R. Lupoi, M. Sparkes, W. O'Neill, Performance comparison of Stellite 6 deposited on steel using supersonic laser deposition and laser cladding. Surf. Coat. Technol. **212**, 119–127 (2012)
32. I. Hemmati, V. Ocelik, K. Csach, J.T.M. de Hosson, Microstructure and phase formation in a rapidly solidified laser-deposited Ni-Cr-B-Si-C hardfacing alloy. Metall. Mater. Trans. A Phys. Metall. Mater. Sci. **45**, 878–892 (2014).

Chapter 5
Supersonic Laser Deposition of Composites

Abstract Supersonic laser deposition (SLD) process as a hybrid coating technique takes the advantages of both cold spray (CS) and laser irradiation, becoming a relatively new trend in the field of coating deposition. This technique has been employed to produce diamond/Ni60 composite and WC-reinforced composite coatings. Two different sizes of diamond particles mixing with Ni60 alloy powder are deposited on medium carbon steel (AISI 1045 steel) substrate with the same optimal parameters, respectively. The diamond/Ni60 composite coatings prepared by process are studied using SEM/EDS, Raman spectra, Vickers hardness and pin-on-disc wear tests. The results show that the diamond/Ni60 composite coatings have not experienced graphitization of diamond thus exhibit better tribological properties, the smaller size of diamond is better than the larger one in the composite coatings for wear resistance. The WC/SS316L composite coatings produced by CS and SLD are compared, with respect to deposition efficiency, WC distribution and concentration, interfacial bonding, phases in the microstructures and tribological properties. The experimental results demonstrated that deposition efficiency, WC concentration and interface bonding of the composite coating can be improved by laser irradiation due to the softening of both powder particles and substrate. The SLD composite coating has the same phases as the CS coating does because of the relatively low heat involved in this process. The high concentration and strong interfacial bonding of WC particles in the SS316L matrix significantly improve the tribological properties of the SLD composite coating. Similar findings are obtained on the WC/Stellite 6 coating fabricated via SLD, compared with the laser cladding (LC) coating. The SLD WC/Stellite 6 coating appears continuous and dense without obvious defects such as pores and cracks.

J. Yao et al., *Advanced Laser Process for Surface Enhancement*, Advanced Topics in Science and Technology in China 61, https://doi.org/10.1007/978-981-15-9659-9_5

5.1 Supersonic Laser Deposition of Diamond/Ni60 Composite Coating

5.1.1 Diamond/Ni60 Composite Coating

Diamond is the hardest material and possesses the highest thermal conductivity [1]. The unique properties of diamond make it the best candidate for cutting and wear-resistant applications. Currently, the main methods used to produce diamond coatings on polymer or metal substrate include laser cladding, thermal spraying, chemical vapor deposition (CVD) and detonation spray, etc. However, it was demonstrated that the heat generated in the thermal spray process can cause graphitization and oxidation of the sprayed diamond in atmosphere environment, leading to weak bond strength and high porosity of the coating [2]. Laser cladding (LC) is often used to produce metal matrix diamond coatings, but the thermal degradation of the diamond is a very serious problem, thus mixture of diamond and matrix in laser cladding process is hard to develop a strong chemical bonding and enough wetting between diamond particles and the matrix [3]. Therefore, effectively controlling the deposition temperature and atmosphere in the LC process to achieve good diamond or diamond composite coatings is a potential tendency. On the other hand, cold spray (CS), owing to low temperature feature, is mainly employed to deposit low hardness, heat sensitive and oxidation sensitive materials. The mixed powder of soft matrix metal and hard diamond particles had been used to fabricate diamond composite coatings by CS, but the poor mechanical properties caused by lower-hardness binding phase have restricted industrialized applications of diamond composite coatings [3].

Ni-based alloys, for example, Ni60 alloy, possess good wettability with diamond and high hardness, which provides diamond with the matrices a larger holding force coefficient to avoid diamond shedding, but depositing such hard materials by CS appears deficient due to limited plastic deformation of the coating and substrate materials. As discussed previously, supersonic laser deposition (SLD), combining the advantages of CS and laser technology, can be the effective means to deposit diamond/Ni60 composite on medium carbon steel substrate [4].

5.1.2 Deposition Process for Diamond/Ni60 Composite Coating

Commercially available artificial diamond powder and Ni60 alloy powder having average size 20 – 25 μm and 10 – 50 μm, respectively, were used as the feedstock materials to create diamond/Ni60 composite coatings. The diamond powder appears irregular, as shown in Fig. 5.1. The composite powder was prepared with 20 wt% diamond and 80 wt% Ni60 powder, which was mechanically milled in the container rotating at a speed of 150 rpm for 2 h in alcohol environment. Stainless steel balls

(a)

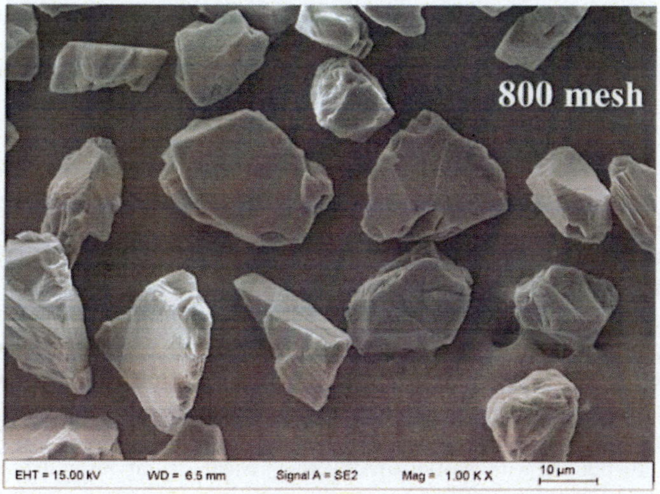

(b)

Fig. 5.1 Morphology of diamond powder having **a** larger particle and **b** smaller particle size [4] with 8 or 12 mm diameter were used as the grinding media and the ball-to-powder weight ratio was 30: 1. The substrate material is AISI 1045 medium steel and the specimen has the dimensions of 100 × 60 × 10 mm. The specimen surface was blasted using 24 mesh Al_2O_3, then ultrasonically cleaned in alcohol medium before SLD processing.

The SLD system used and the SLD experimental procedures in this study are the same as those used to fabricate the Stellite 6 and Ni60 alloy coatings, as described previously. Because of the softened substrate and powder particles by means of laser irradiation, the diamond/Ni60 composite coating can be formed at a very reduced impact velocity (about half of that used in CS). The nitrogen was used to accelerate the composite powder. The laser spot diameter was 5 mm. The spraying distance was 25 mm; the traverse rate was 30 mm/s, the nitrogen temperature was 550 °C, the nitrogen gas pressure was 30 bar.

5.1.3 Microstructure of Diamond/Ni60 Composite Coating

The cross-section of each coating specimen was examined using SEM, to analyze the deposition efficiency, diamond particle distribution and microstructure. The coating layer thickness and area ratio of diamond particles of the specimens were measured with the image analysis software of SEM (ΣIGMA HV-01–043, Carl Zeiss SMT Pte Ltd, Germany). The thermal degradation, graphitization and oxidation of diamond particles were investigated by X-ray diffraction (XRD, Bruker D8 Advance, Bruker, Germany) and Raman spectrum (Lab RAM HR UV800, JOBIN YVON, France), respectively. As shown in Fig. 5.2a and b, whether the coating made from larger-size diamond particles or from small-size particles appears bonded mechanically with the substrate, but the former shows denser interface than the latter. The micro-indentation was performed at the interface of the coatings, as shown in Fig. 5.2c and d. It can be seen that under the maximum indentation load of 9.8 N (machine capacity) no cracks are found around the indentation marks, which indicates good interface bonding of the coatings. Additionally, the microstructures of both coatings show uniform diamond particle distribution within the Ni60 matrix in Fig. 5.2e and f. At high magnification of SEM images, it is observed that diamond particles are solidly bonded with Ni60 matrix by mechanical locking for both coatings, as shown in Fig. 5.2g and h.

Ni60 alloy has lower melting point, higher hardness and better compatibility with diamond, compared with Cu, Al and Ti, which suggests that it is a better binding matrix of diamond. The volume fractions of diamond in the diamond/Ni60 composite coatings were estimated using the image analysis software of SEM and are indicated in Figs. 5.3a and b, where the dark area represents diamond particles. In general, the two coatings have similar volume fraction of diamond. In high temperature and oxidation environments diamond is likely to burn or graphitize, but no graphitization of diamond particles was found in the two coatings. This is attributed to low deposition temperature and inert nitrogen gas used in the SLD process, compared with the

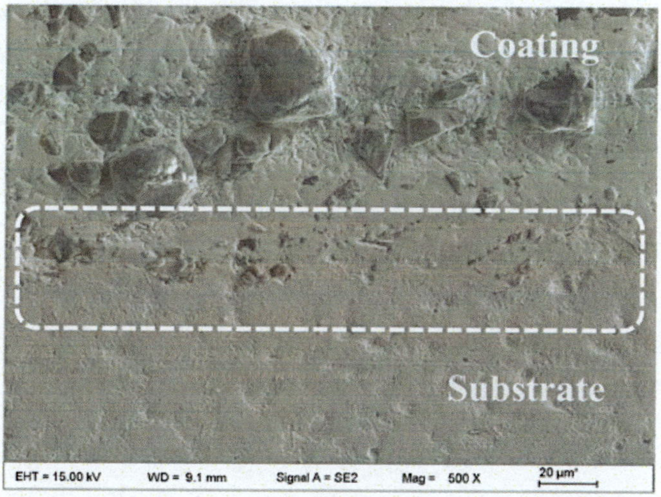

(a)

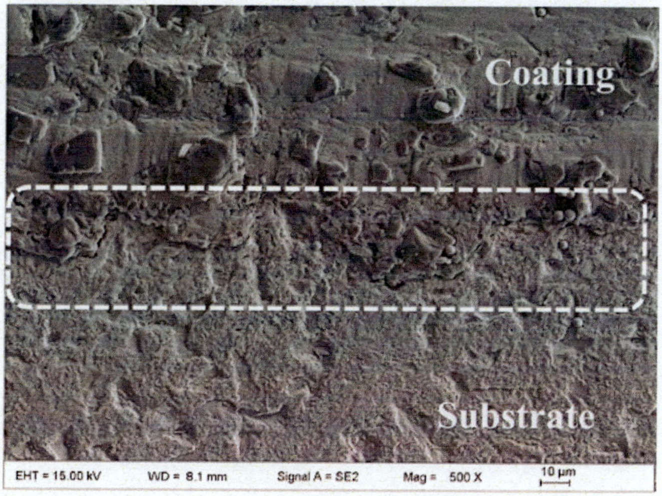

(b)

Fig. 5.2 SEM microstructure of diamond/Ni60 composite coating: **a** cross-section of the coating with large-size diamond particles, **b** cross-section of the coating with small-size diamond particles, **c** indentation test on the coating with large-size diamond particles, **d** indentation test on the coating with small-size diamond particles, **e** diamond particle distribution in the coating with large-size diamond particles, **f** diamond particle distribution in the coating with small-size diamond particles, **g** interfacial bonding between diamond particles and Ni60 matrix in the coating with large-size diamond particles and **h** interfacial bonding between diamond particles and Ni60 matrix in the coating with small-size diamond particles [4]

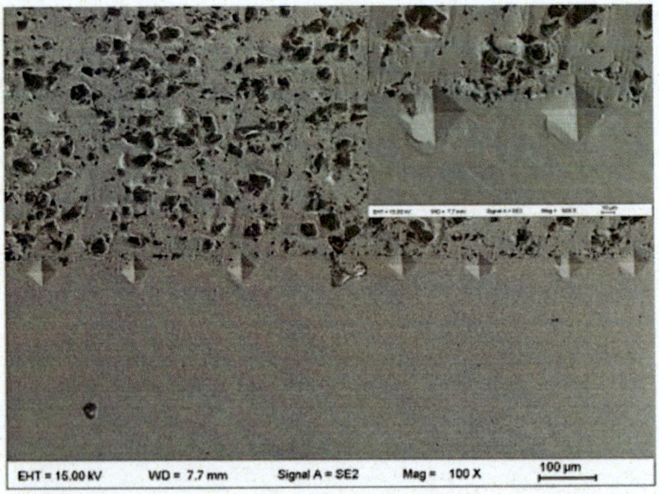

(c)

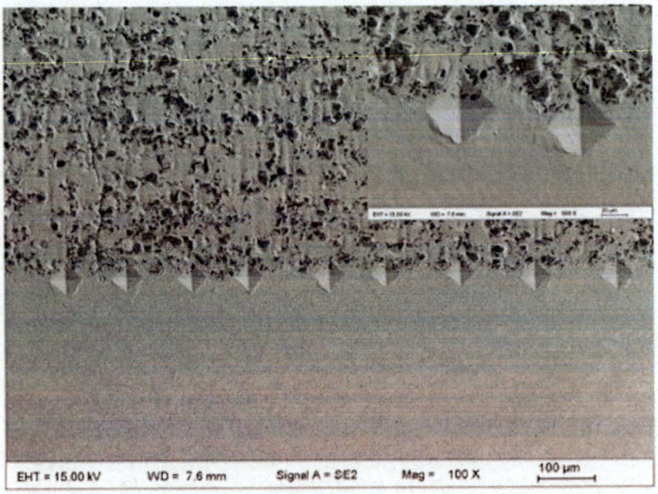

(d)

Fig. 5.2 (continued)

thermal spray and laser cladding process. To further investigate diamond graphitization behavior of the diamond/Ni60 composite coatings, Raman spectra analysis was made on the coatings using a laser source with a wavelength of 632.81 nm and a spot of 300 µm. A Raman shift ranging from 500 to 2500 cm^{-1} was employed with a dwelling time of 10 s. As illustrated in Fig. 5.3c, a single and sharp peak of diamond

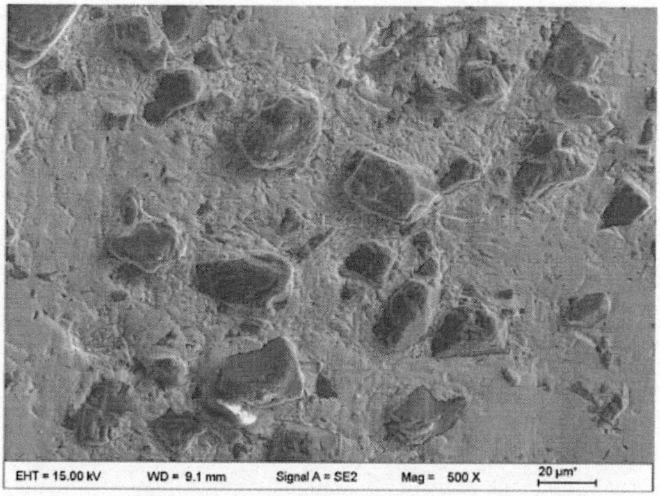

(e)

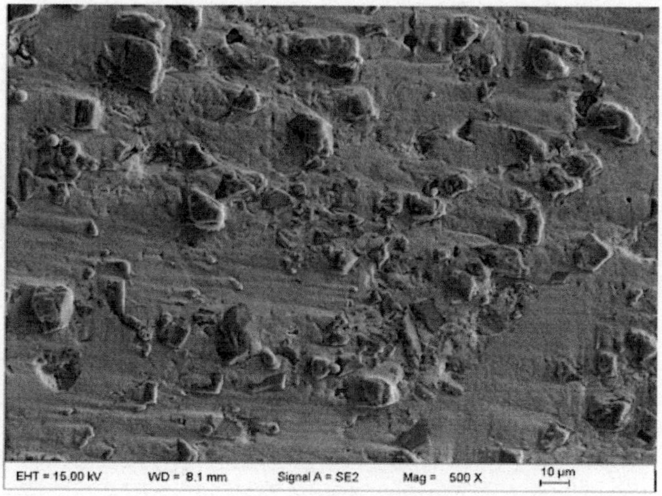

(f)

Fig. 5.2 (continued)

is observed at 1339 cm^{-1} for both coatings, which matches the diamond peak of feedstock diamond powder. This reveals that diamond particles were not graphitized in the SLD process when fabricating the diamond/Ni60 composite coatings.

X-ray diffraction (XRD) analysis was conducted on the diamond/Ni60 composite coatings prepared at the deposition temperature of 980 °C to identify the phases

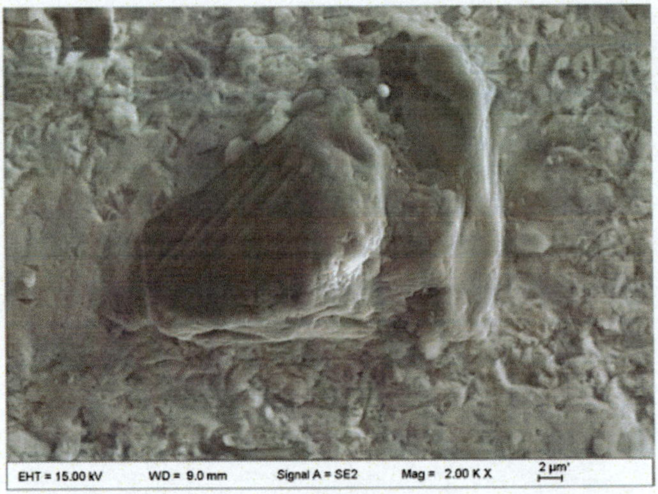

(g)

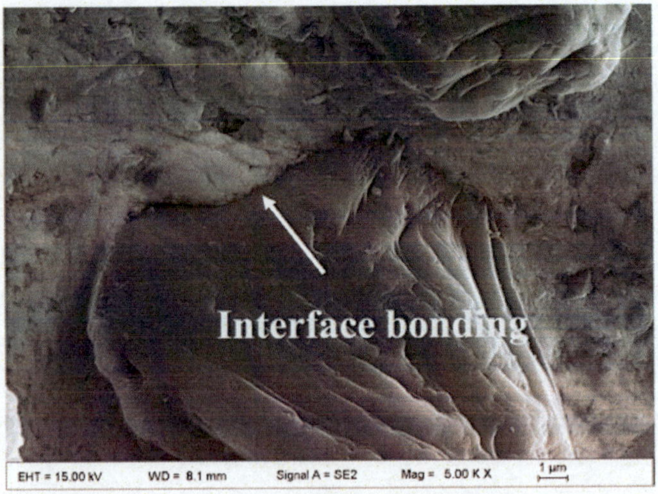

(h)

Fig. 5.2 (continued)

formed in the microstructures. The analysis was operated at 40 kV and 40 mA to generate monochromatic Cu K_α radiation with a wavelength of 0.154 nm, and a 2θ diffraction angle ranging from 30° to 80° was employed with a time interval of 5 s and a step size 0.05°. As presented in Fig. 5.4, the phases present in the coatings include γ-Ni, FeNi and Cr_mC_n. The compounds Ni_mSi_n and FeNi are the main phases

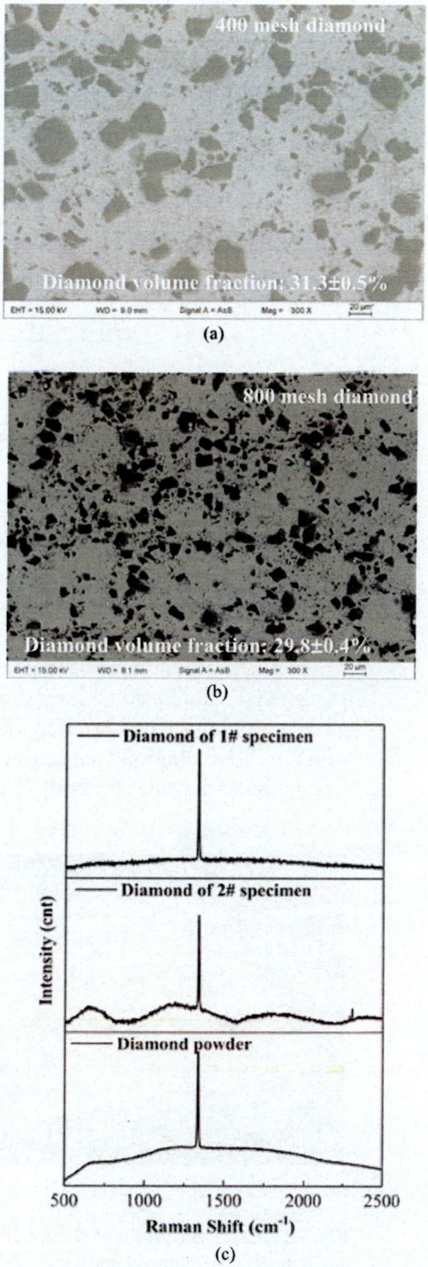

Fig. 5.3 Diamond graphitization analyses: **a** volume fraction of diamond in the coating with large-size diamond particles, **b** volume fraction of diamond in the coating with small-size diamond particles and **c** Raman spectra [4]

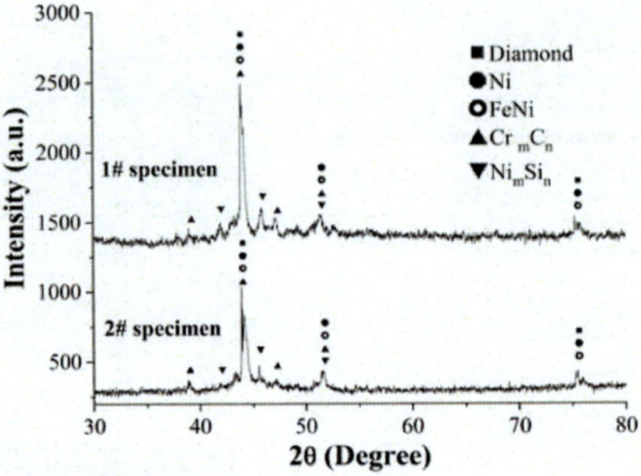

Fig. 5.4 X-ray diffraction patterns of diamond/Ni60 composite coatings [4]

of the solid solution of Ni60 alloy, and the intermetallic Cr_mC_n and Ni_mSi_n phases consist of Cr_7C_3, $Cr_{23}C_6$, Ni_3Si_2, and Ni_3Si, which are the strengthening phases of Ni60 alloy. This further confirms that the diamond/Ni60 composite coatings prepared via SLD retain the microstructures of feedstock materials, exhibiting solidus-state deposition characteristics.

The surface roughness of the diamond/Ni60 composite coatings was investigated using a three-dimensional laser microscopy (LEXT OLS4100, OLYMPUS) and the surface profiles are presented in Fig. 5.5. The mean surface roughness of the coating with large-size diamond particles is about 90 μm and that of the coating with small-size diamond particles is about 75 μm, that is, the latter is smoother than the former.

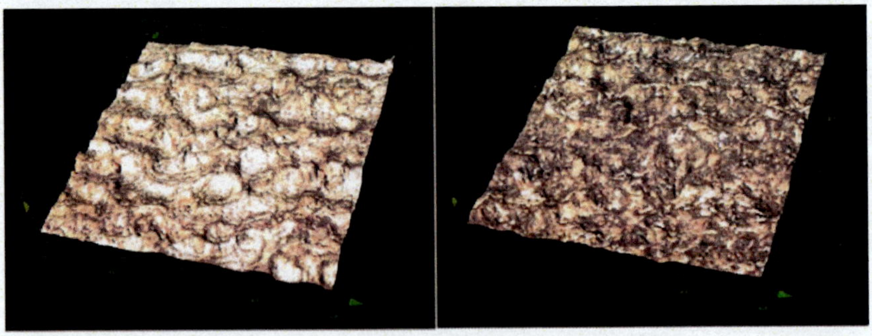

(a) (b)

Fig. 5.5 3D surface profile of diamond/Ni60 composite coating with **a** large-size diamond particles and **b** small-size diamond particles [4]

5.1.4 *Hardness and Tribological Performance of Diamond/Ni60 Composite Coating*

The hardness of the Ni60 matrix of the diamond/Ni60 composite coatings was measured using a Digital Micro-Hardmeter (HMV-2 T, SHIMADZU), as demonstrated in Fig. 5.6a, along with the hardness measurement on pure Ni60 coating produced via SLD for comparison. The average hardness results are reported in

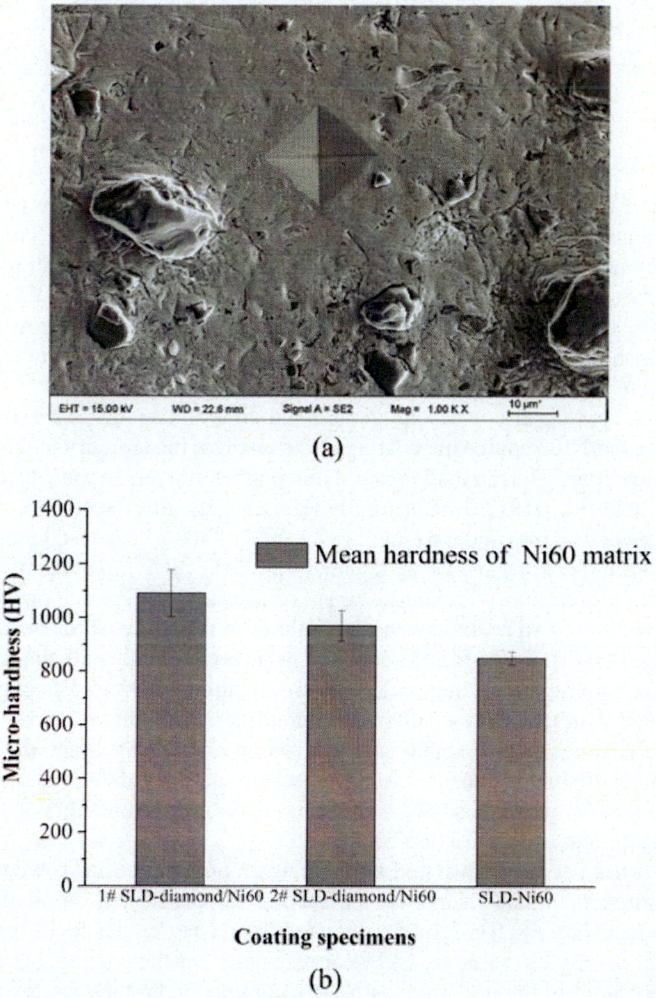

Fig. 5.6 Microhardness test on Ni60 matrix: **a** indentation on Ni60 matrix and **b** average hardness results [4]

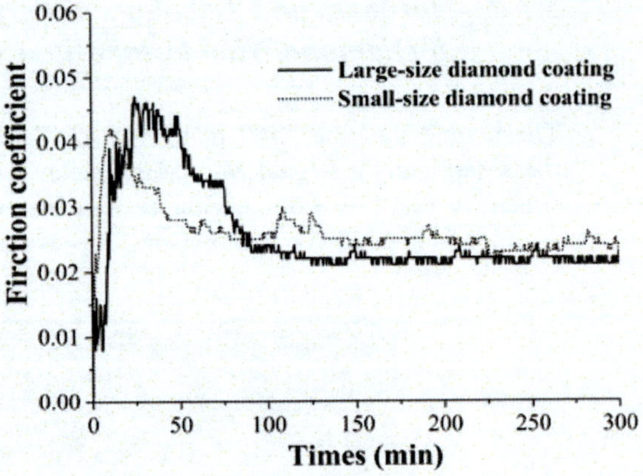

Fig. 5.7 Friction coefficient evolution with sliding time of diamond/Ni60 composite coatings [4]

Fig. 5.6b. It is evident that the Ni60 matrix of the composite coatings is harder than pure Ni60 coating and the Ni60 matrix of the composite coating with large-size diamond particles is harder than that of the composite coating with small-size diamond particles. The difference in hardness of Ni60 alloy in these three coatings suggests that although the SLD process can keep the phases of feedstock Ni60 powder unchanged, the ratio of the strengthening phases in Ni60 alloy may vary among the coatings, thus resulting in hardness difference of Ni60 matrix.

A pin-on-disc wear test was used to evaluate the wear resistance of the diamond/Ni60 composite coatings. The test was made in dry-sliding mode at room temperature. The pin (ball) was a 4 mm diameter Si_3N_4 ceramic ball with a hardness of 2200 HV (HRC90). During the wear test, the disc (specimen) was rotated at a speed of 500 rpm under a compressive load of 500 g. The test duration was 300 min for each specimen and the generated wear track was 5 mm in diameter. The recorded friction evolution with sliding time is illustrated in Fig. 5.7 for both coatings. In general, the two coatings behave similarly in friction coefficient. During the initial stage of 100 min, the friction coefficients increase rapidly until reaching a maximum. At this point, the coating with large-size diamond particles has larger friction coefficient than that with small-size diamond particles, but with wear proceeding, the two coatings have a stable smaller friction coefficient. Since the hardness of the diamond/Ni60 composite coatings is governed by the amount of diamond present in the coating, while these two coatings have similar volume fraction of diamond, thus they exhibit similar friction behavior.

It has been demonstrated that diamond/Ni60 composite powder with both hard reinforcement and matrix can be successfully deposited on medium carbon steel via SLD process. The particle size of diamond can influence the microstructure and final properties (hardness and wear resistance) of the coating, but not significantly. According to the principles of SLD technique, definitely, large-size hard particles require higher energy, including higher heat power and higher impact velocity, etc., in order to effectively deposit the particles on the substrate. In other words, with the same process parameters, small-size hard particles can easier achieve denser and well-bonded coating from SLD process. Therefore, it is suggested that for diamond/Ni60 composite coating, adding the same weight percentage of diamond small-size particles would be better for creating a good coating with strong interfacial bonding by mechanical locking.

5.2 Supersonic Laser Deposition of WC/Stainless Steel Coatings

5.2.1 Tungsten-Carbide-Reinforced Coatings

Tungsten carbide (WC) is one of the common reinforcing agents used to create ceramic/metal composites for wear resistance applications, owing to its high melting point, high hardness and strength [5]. As discussed above, since supersonic laser deposition (SLD) technique can reduce particle deposition temperature and critical deposition velocity by introducing laser illumination into cold spray (CS), hard particles, for example, Stellite alloy and Ni60 alloy powder can be deposited at a lower impact velocity for high coating density and strong interfacial bonding. Meanwhile, the low-temperature deposition feature of SLD can effectively avoid high thermal stress, oxidation, phase transformation and grain growth of the coatings. Further study in SLD was made on creating WC-reinforced stainless steel composite coatings on stainless steel substrate [6].

5.2.2 Deposition Process of WC/SS316L Composite Coatings

Commercially available stainless steel 316L (SS316L) powder and WC powder were used as feedstock materials for creating WC/316L coatings. The particle shapes of SS316L powder and WC powder are spherical and irregular, respectively, as shown in Fig. 5.8, along with the SS316L particle size distributions illustrated.

The composite powder consisting of 30 vol% WC and 70 vol% SS316L was mechanically milled in a cylinder rotating at a speed of 200 rpm for a time period of 2 h to obtain a uniform powder mixture. Stainless steel balls having 8 or 12 mm diameter were used as the grinding media and the ball-to-powder weight ratio was 30: 1. Because of low milling speed and short milling time used, the WC particles did not stick to the stainless steel particles during the ball milling process, thus the morphology and size of the WC particles were not affected by the ball milling process, as seen in Fig. 5.8d. The substrate material was stainless steel and the substrate specimens had a dimension of $100 \times 60 \times 10$ mm. Before the coating process, the substrate surface was grit-blasted using 24 mesh alumina and ultrasonically cleaned in alcohol.

Fig. 5.8 Composite coating feedstock materials **a** morphology of SS316L powder, **b** morphology of WC, **c** size distribution of SS316L particles and **d** morphology of composite powder after milling [6]

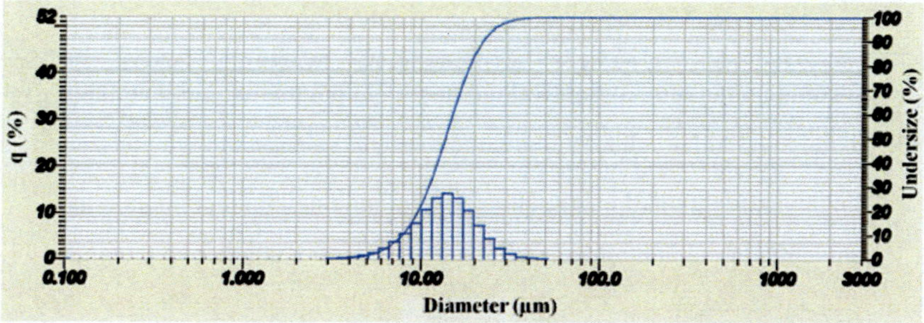

(c)

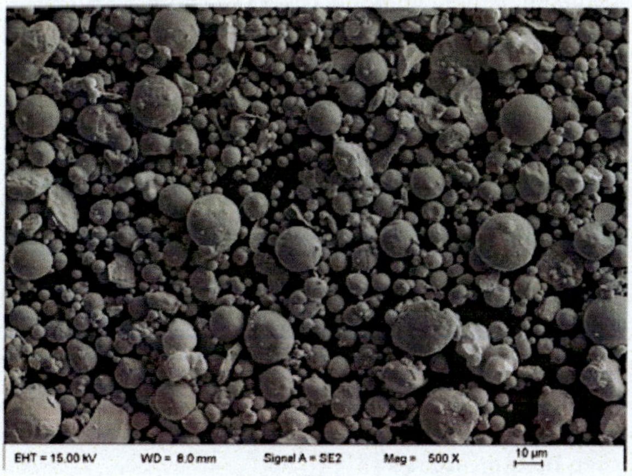

(d)

Fig. 5.8 (continued)

The SLD system used and the SLD experimental procedures in this study are the same as those used to fabricate the Stellite 6 and Ni60 alloy coatings, as described previously. Combined lenses were used to focus the laser beam onto the substrate surface and the diameter of the laser spot was 5 mm. The working gas was nitrogen gas. The detailed process parameters of coating production are reported in Table 5.1.

5.2.3 Deposition Efficiency of WC/SS316L Composite Coatings

The specimens were prepared following conventional metallography procedures and then etched with an aqueous solution consisting of 45 mL HCl, 15 mL HNO_3, and

Table 5.1 Process parameters for fabricating WC/SS316L composite coatings [6]

Sample number	Gas pressure (MPa)	Gas temperature (°C)	Deposition zone temperature (°C)	Laser traverse speed (mm/s)	Powder feeding rate (g/min)	Spraying distance (mm)
a	3	450	Ambient temperature	10	40	30
b	3	450	500	10	40	30
c	3	450	700	10	40	30
d	3	450	900	10	40	30

20 mL methanol. A field emission SEM (ΣIGMA HV-01–043, Carl Zeiss) coupled with EDS (Nano Xflash Detector 5010, Bruker) was used in the microstructural analyses.

From the coating cross-section, the coating layer thickness can be measured under SEM, thus the deposition efficiency of the SLD process can be determined. All coatings are single layer and the track offset for large area of coating is 50% of the track width. As shown in Fig. 5.9, since the laser spot (5 mm diameter) is smaller than the powder beam (8 mm diameter) and the distribution of laser energy is non-uniform with a near tophat intensity distribution, the coatings have a variable thickness along cross-section with maximum peak occurring at the center of cross-section. Also, it is observed that maximum height strongly depends on deposition temperature and increases from 869.5 μm to 1.153 mm when deposition temperature is increased from room temperature to 900 °C.

A key concept in CS process is critical velocity. Theoretical modeling of critical velocity can be expressed as [7]

$$V_{\mathrm{cr}} = 667 - 14\rho + 0.08T_{\mathrm{m}} + 0.1\sigma_{\mathrm{u}} - 0.4T_{\mathrm{i}} \tag{5.1}$$

where ρ is the density of the material, T_{m} is the melting temperature, σ_u is the ultimate strength, and T_{i} is the initial particle temperature. According to Eq. (5.1), particle preheating will decrease the critical velocity because as the initial particle temperature is increased and the ultimate strength of the materials is reduced. In this study, the spraying nozzle was perpendicular to the substrate surface and the laser beam was at an angle of 30° to the surface normal. The powder stream and the laser beam partially overlapped with each other. Although the spraying particles were travelling at high velocities and have limited time of exposure to the laser, it was expected that the particles were significantly heated in flight by laser prior to impacting the substrate because of high laser power density and small particle size, which could reduce the critical velocity of spraying particles. As a result, the proportion of particles exceeding this velocity would increase, leading to the improvement in deposition efficiency, as shown in Fig. 5.9. In addition, the heated particles easier deformed and adhered to the substrate and the previously deposited particles to form the coating, thus also improving deposition efficiency.

5.2.4 Microstructure of WC/SS316L Composite Coatings

The WC concentration and distribution in the WC/SS316L composite coatings were investigated using SEM associated image analysis software. It is found that WC

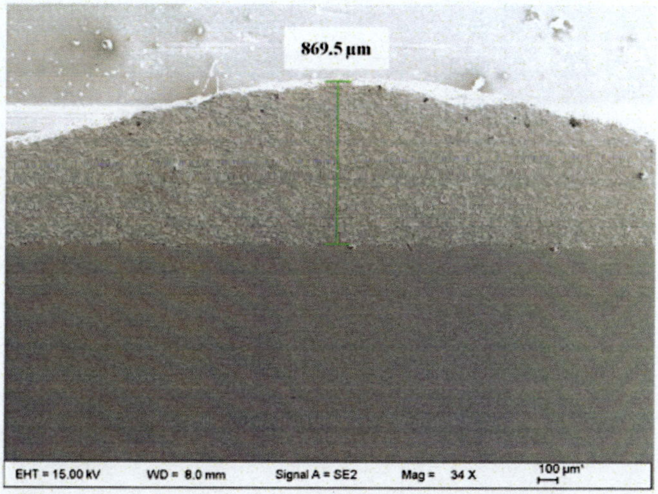

(a)

(b)

Fig. 5.9 SEM image of cross-section of WC/SS316L composite coating prepared under the deposition temperature of **a** room temperature, **b** 500 °C, **c** 700 °C and **d** 900 °C [6]

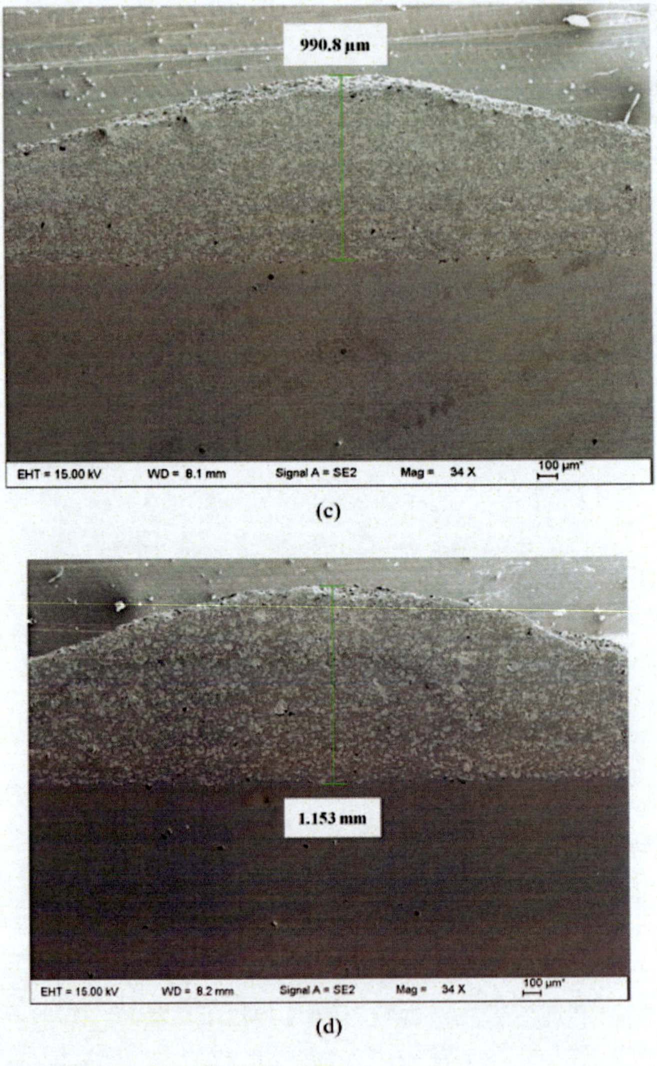

Fig. 5.9 (continued)

particles are uniformly distributed in all coatings and the concentration of WC particles in the coatings is increased by raising deposition temperature (Fig. 5.10). The WC volume fractions in the coatings were estimated and the results are illustrated in Fig. 5.11. It is shown that the volume fraction of WC particles in the coatings does not change significantly when deposition temperature increases from ambient temperature to 500 °C, but it increases greatly when deposition temperature reaches 700 °C or above. The coating deposited at 900 °C has about 29.3% volume fraction

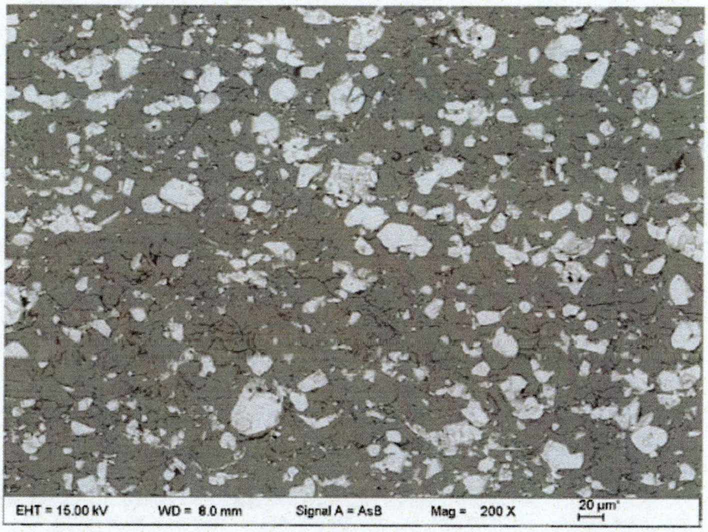

(a)

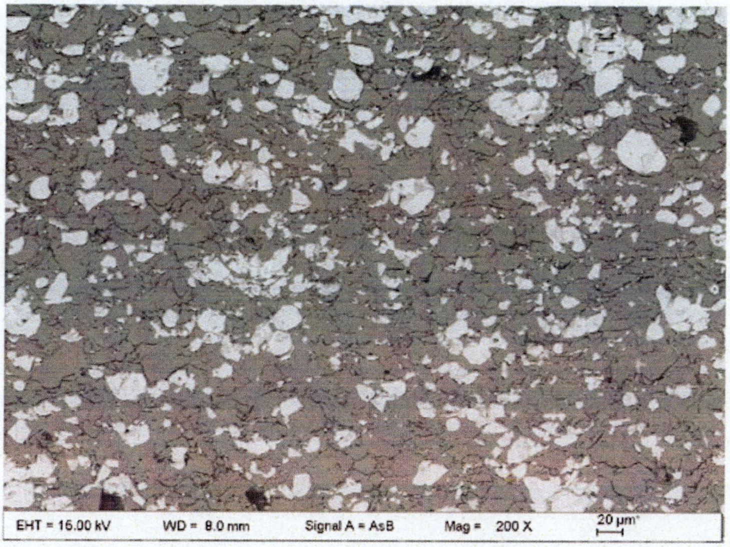

(b)

Fig. 5.10 Distribution of WC particles in WC/SS316L coating prepared under the deposition temperature of **a** room temperature, **b** 500 °C, **c** 700 °C and **d** 900 °C [6]

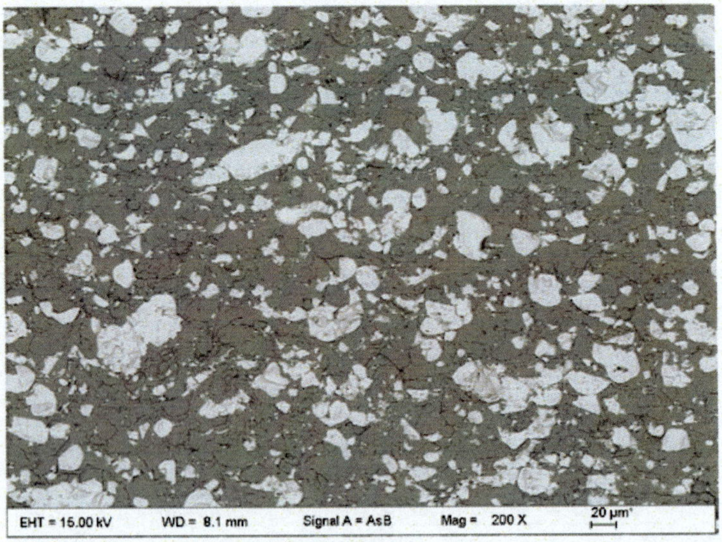

(c)

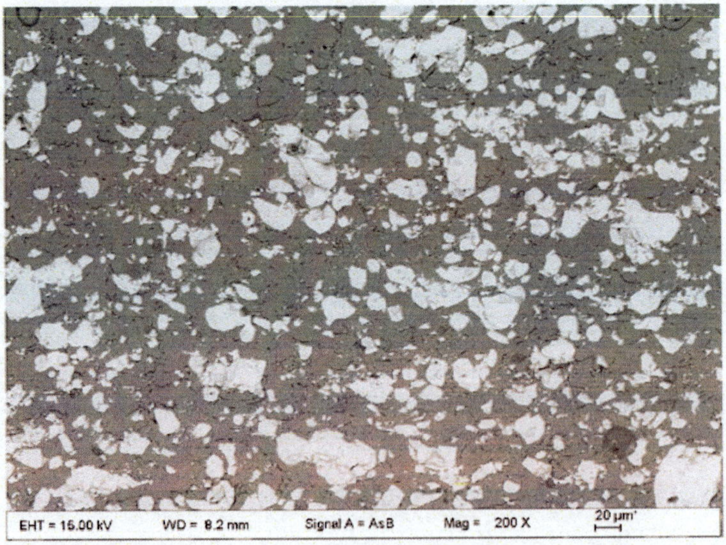

(d)

Fig. 5.10 (continued)

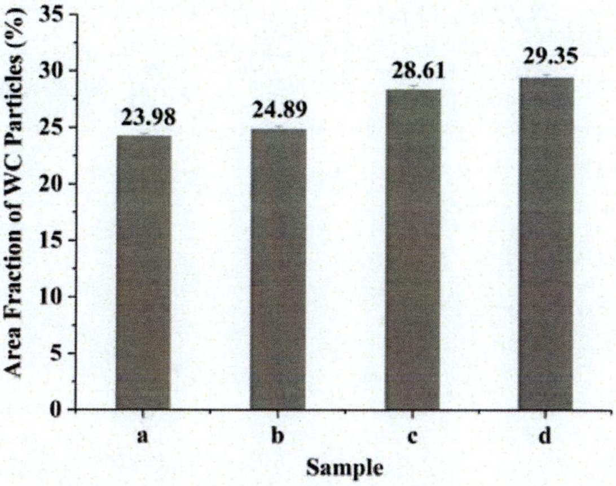

Fig. 5.11 Illustration of WC volume fraction in WC/SS316L composite coatings [6]

of WC particles, which is very close to the volume fraction (30%) of WC particles in the composite powder. This means that laser heating of the powder and substrate benefits deposition of WC particles in the coatings.

The phases in the feedstock powders and in the composite coatings were investigated using an X-ray diffractometer (XRD, D8 Advance, Brucker) with Cu-$K\alpha$ radiation, 45 kV, 40 mA, at a scan rate of 0.02°/s. As presented in Fig. 5.12, the XRD diffraction patterns reveal that the feedstock powder (#1 specimen) and the composite coating (#2 specimen) have the same phases, that is, they all have the phase of SS316L and WC. Also, the phases of the coatings produced by CS and by SLD are identical. EDS analysis was performed on the SS316L matrix of the CS and SLD composite coatings and the results are illustrated in Fig. 5.13. The oxygen element content is trivial in both coatings, which confirms that SLD is a solidus-state particle deposition process like CS although laser power provides heat energy on the

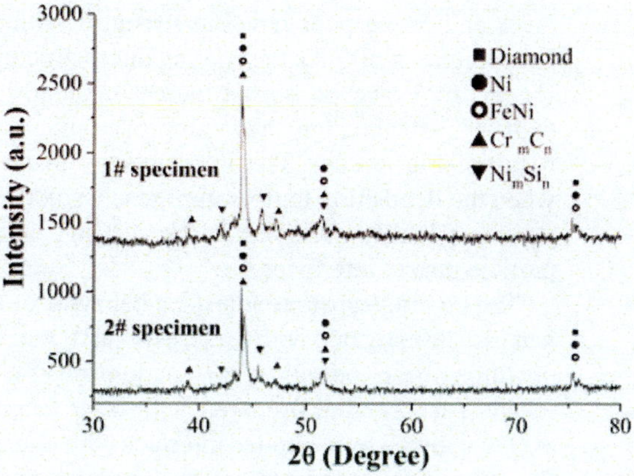

Fig. 5.12 XRD diffraction patterns of WC/SS316L composite powder and coating [6]

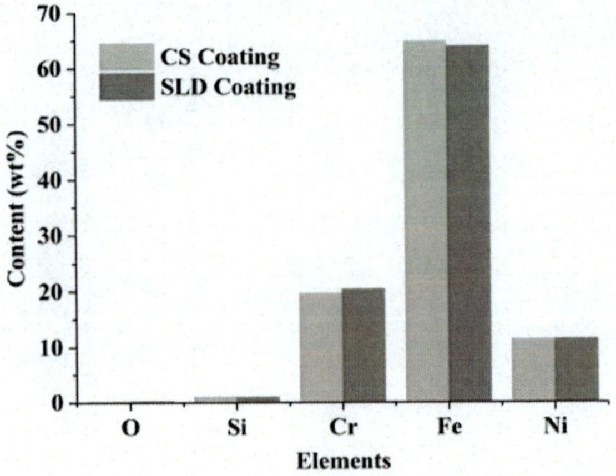

Fig. 5.13 Element concentrations of WC/SS316L composite coatings [6]

coating and substrate materials. The contents of other elements in the CS and SLD coatings are similar. This is because in the SLD process laser power was elaborately controlled to only soften the coating and substrate materials, but not to melt them. In this case, the particles remained in solidus state during flight and deposition, and they were subject to laser irradiation for only a limited period of time, thus the chemical composition and phases of the coating would not be affected by laser heating.

5.2.5 Interface Bonding of WC/SS316L Composite Coatings

Further study in beneficial effect of laser corporation in CS on producing WC/SS316L composite coating was made by examining the interfacial bonding between the WC particles and SS316L matrix using SEM. As seen in Fig. 5.14a, without laser heat applied, the coating contains lots of gaps and pores at the particle/matrix interfaces, indicating poor bonding strength. With laser heat applied, at the deposition temperature of 500 °C, the coating microstructure appears not improved, as shown in Fig. 5.14b. However, further increasing deposition temperature by laser heating up to 700 °C, the coating shows improved microstructure with gaps and pores greatly reduced (Fig. 5.14c). The improvement in the coating microstructure is apparent when the deposition temperature reaches 900 °C, as shown in Fig. 5.14d, gaps and pores are hardly observed in the coating, indicating good bonding formed at the particle/matrix interfaces.

The coating/substrate interface behavior of the WC/SS316L composite coatings was also investigated with SEM. Similarly, without the assistance of laser heating, the coating appears porous at the coating/substrate interfaces, as shown in Fig. 5.15a, whereas the coating prepared with laser heating at the deposition temperature of 900 °C shows a well-bonded interface with minimum porosity (Fig. 5.15b). Currently, the most widely accepted bonding mechanisms of coatings by CS is adiabatic shear

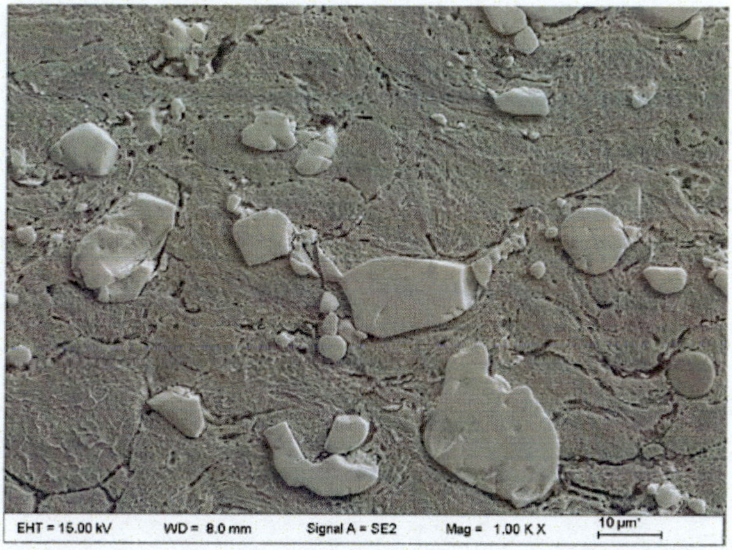

(a)

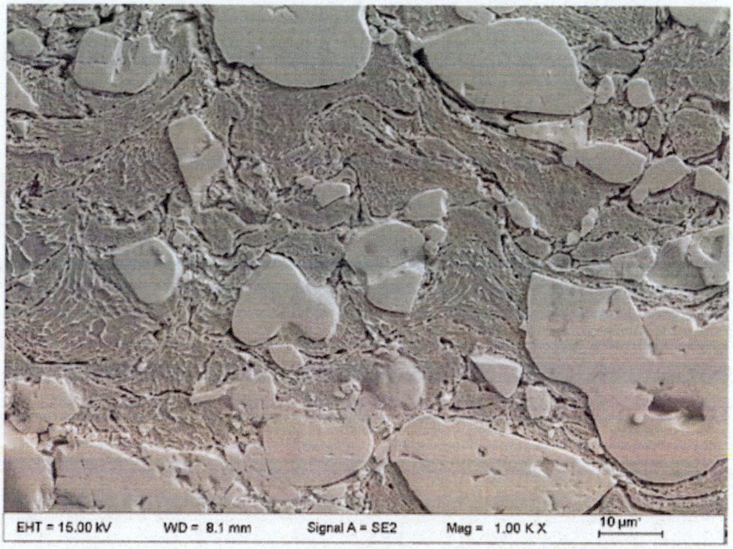

(b)

Fig. 5.14 SEM image showing particle/matrix interface bonding of WC/SS316L coating prepared under the deposition temperature of **a** room temperature, **b** 500 °C, **c** 700 °C and **d** 900 °C [6]

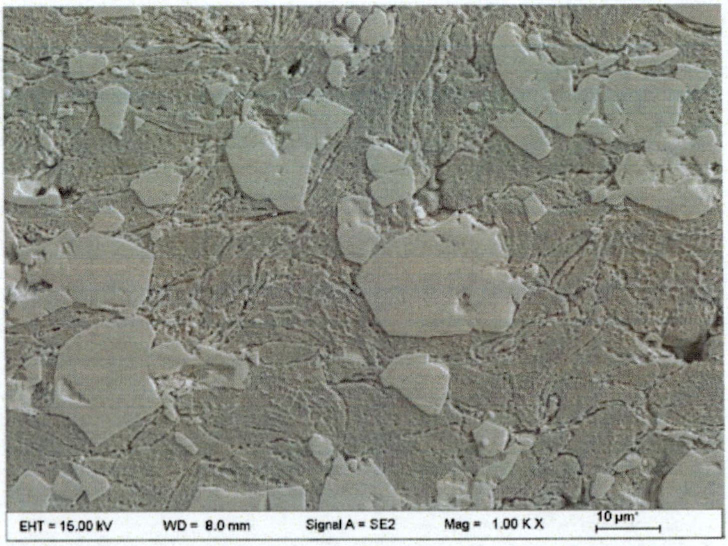

(c)

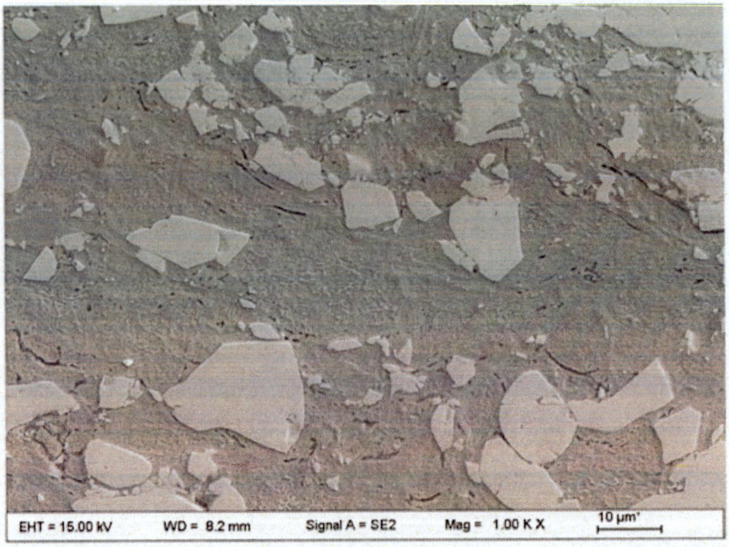

(d)

Fig. 5.14 (continued)

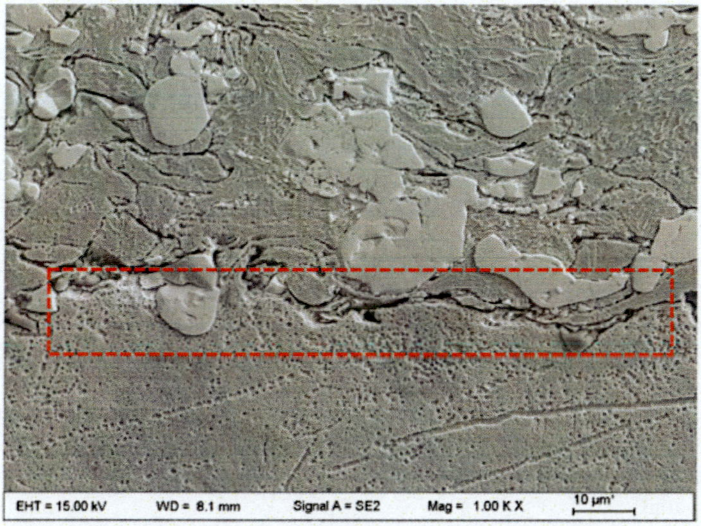

(a)

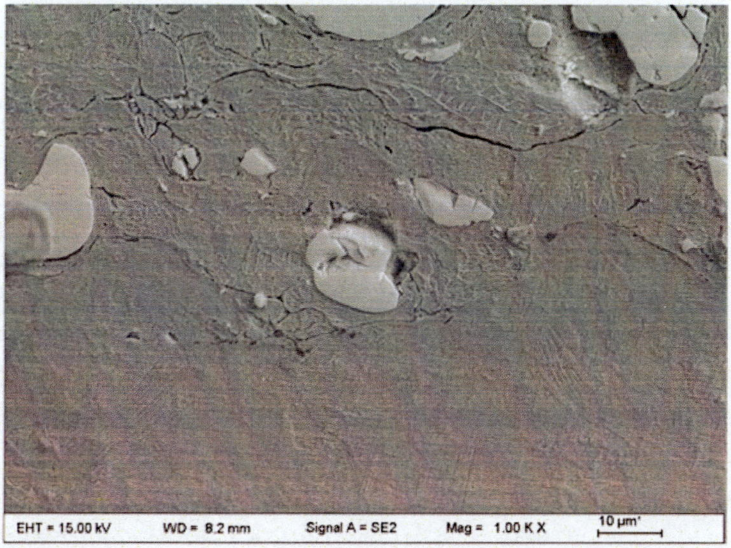

(b)

Fig. 5.15 SEM image showing the coating/substrate interface bonding of WC/SS316L coating prepared under the deposition temperature of **a** room temperature and **b** 900 °C [6] instability at the interface, which results from the high strain rate and intensive localized deformation of the coating and substrate materials during the deposition process [7]. Interfacial instability in the high-velocity impacting process of CS causes material roll-ups and vortices at the interface, consequently, the substrate and coating materials merge at the interface region, resulting in mechanical interlocking [8]. Another proposed bonding mechanism of CS is the metallurgical bonding by atomic diffusion between coating and substrate materials, which can provide better bonding strength compared to mechanical interlocking [9]. However, in CS solidus-state diffusion is insufficient to form a thick diffusion layer due to extremely short timescale of particle–substrate interaction and low temperature [8]. With the assistance of laser irradiation on the deposition site, the substrate is heated up becoming softer, which favors lodging impacting particles thus forming intimate mechanical interlocking. Moreover, the synchronous laser heating on the deposition site can also promote the atomic diffusion between the coating and substrate materials, which greatly increases the possibility of metallurgical bonding. All these contribute to the better coating/substrate interfacial bonding of SLD coatings compared to CS coatings.

5.2.6 Wear Performance of WC/SS316L Composite Coatings

Pin-on-disc wear test was conducted on the SS316L substrate and the WC/SS316L coatings at room temperature under dry-sliding conditions. The pin was a 4 mm diameter Si_3N_4 ceramic ball, having a hardness of 2200 HV (HRC90). The disk was the specimen that was polished, cleaned in an ultrasonic bath, and finally dried. The test was performed under a normal load of 500 g at a rotational speed of 500 rpm of the disc between the contacting surfaces, and the sliding duration was 60 min for each specimen. The friction coefficients recorded during the wear tests are illustrated in Fig. 5.16. The friction coefficient of the substrate is much larger than that of the composite coatings, indicating that the WC/SS316L coatings can significantly improve the wear-resistant properties of SS316L due to the high cutting action of irregular WC particles. Also, as observed, the friction coefficient of the SLD coating is smaller and more stable than that of the CS coating. The average coefficient of the SLD coating is about 0.39 after 60 min sliding, while that of the CS one is about 0.54 at the same time.

The worn surfaces of the WC/SS316L coatings were examined using SEM. As shown in Fig. 5.17, both coatings have two distinct areas in the wear tracks, dark and light, and the CS coating has a more dark area than the SLD coating. EDS analysis was performed on the dark and light area with the results presented in Fig. 5.18. It is shown that the dark area, marked area 1 and area 3 in Fig. 5.17a and b, contain a much higher content of O element than the light area, marked area 2 and area 4 in Fig. 5.17a and b. This indicates that oxidation had occurred in the worn tracks when the coatings were subjected to sliding wear. In the SEM images at high magnification (Fig. 5.17c and d), oxidation debris is found in the worn surfaces. More oxidation

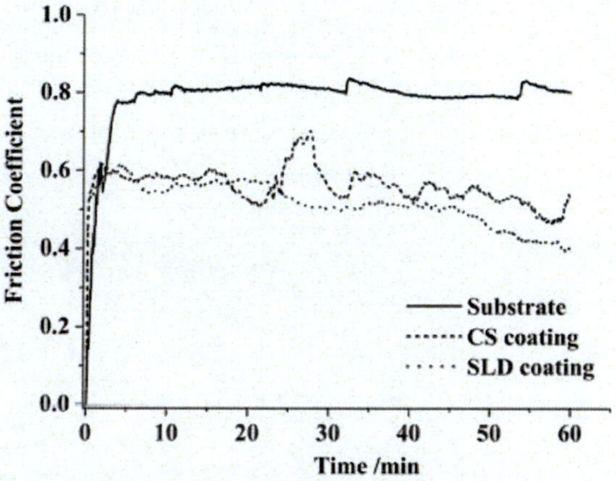

Fig. 5.16 Friction coefficients of WC/SS316L coatings and SS316L substrate recorded in dry-sliding wear tests [6]

debris is found in the worn surface of the CS coating, which is consistent with the friction coefficient observations in Fig. 5.16. Additionally, it is noted in Fig. 5.18 that the light area of the CS coating (area 2) is rich in iron while that of the SLD coating (area 4) tungsten. This can be attributed to higher WC content in the latter.

Based on the wear test results, the WC/SS316L composite coating produced by SLD has better tribological properties than the CS coating. It is believed that the friction between the contact surfaces was reduced due to the abrasion resistance of hard irregular WC particles. The better tribological properties of the SLD coating resulted from the relatively high content of WC particles and better interfacial bonding between WC particles and SS316L matrix owing to the beneficial effects of laser irradiation.

5.3 Supersonic Laser Deposition of WC/Stellite 6 Coatings

5.3.1 Deposition of WC/Stellite 6 Coatings

To extend the application of Stellite 6 alloy in extremely severe wear conditions, WC-reinforced Stellite 6 composite coating was created via SLD [10]. The Stellite 6 powder and WC powder used in this research were commercially available, having a shape spherical and irregular respectively. For comparison, WC/Stellite 6 coating was also prepared via laser cladding (LC). The SLD and LC process required different particle sizes of feedstock powders so that the mean particle sizes of Stellite 6 powder for LC and SLD were 81.87 μm and 14.67 μm respectively and the WC powder had a mean particle size of 32.48 μm for LC and 12.03 μm for SLD. The composite powder consisting of 30 vol% WC and 70 vol% Stellite 6 was mechanically milled in a cylinder rotating at a speed of 200 rpm for 2 h to obtain a uniform mixture.

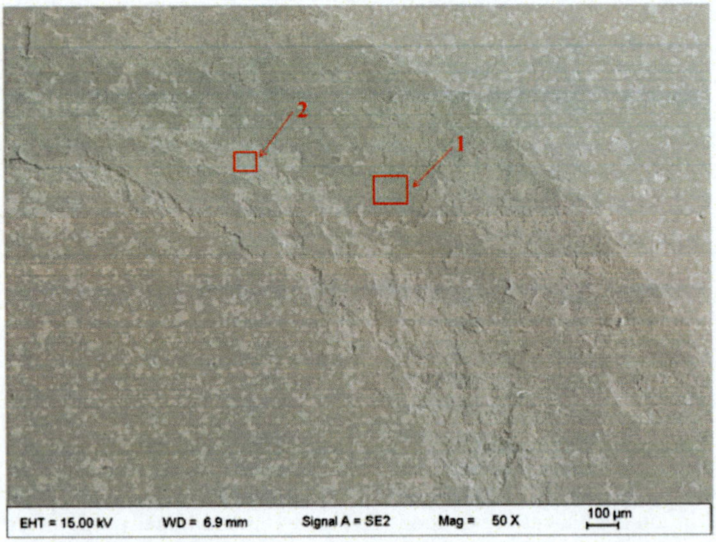

(a)

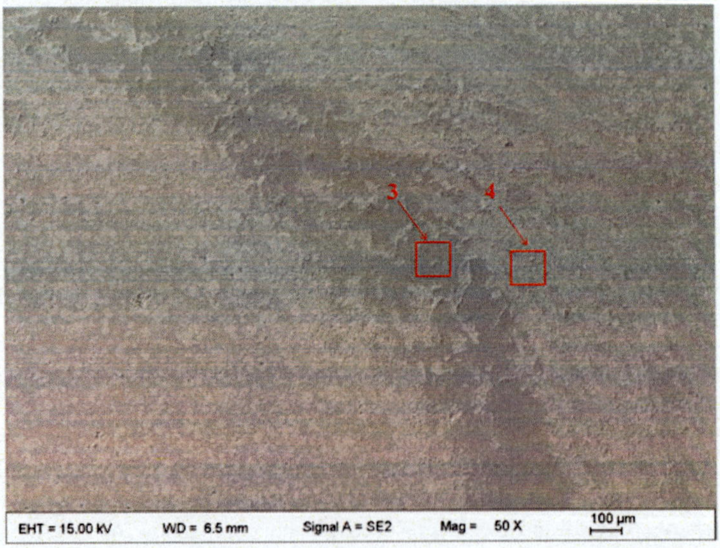

(b)

Fig. 5.17 Worn surface morphology of **a** the CS coating at low magnification, **b** the SLD coating at low magnification, **c** the CS coating at high magnification and **d** the SLD coating at high magnification [6]

(c)

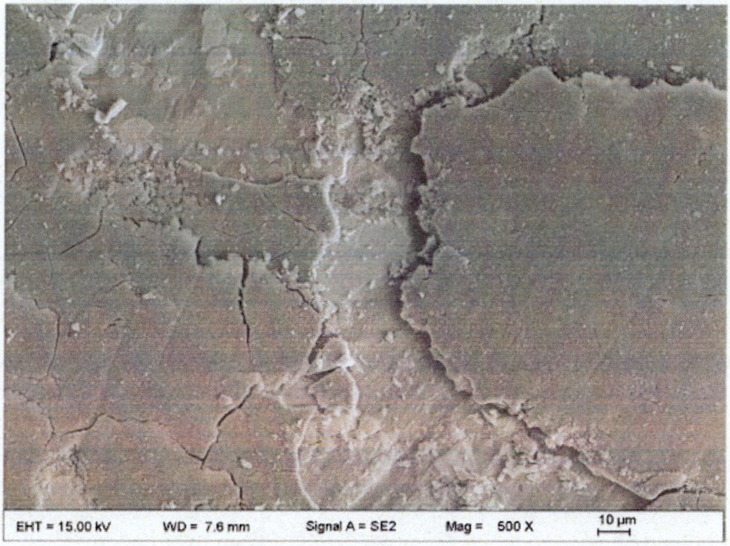

(d)

Fig. 5.17 (continued)

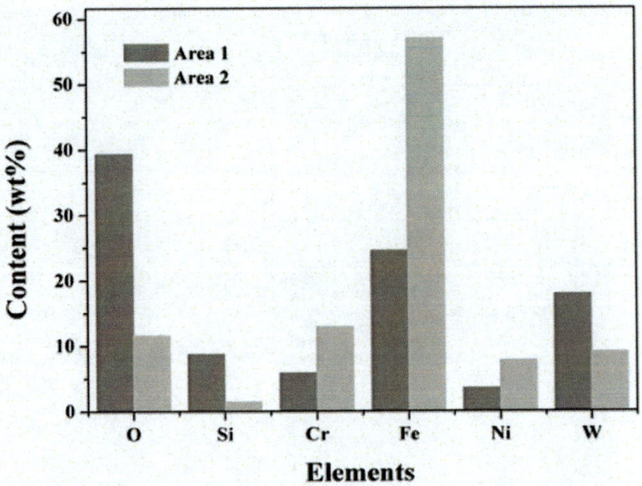

(a)

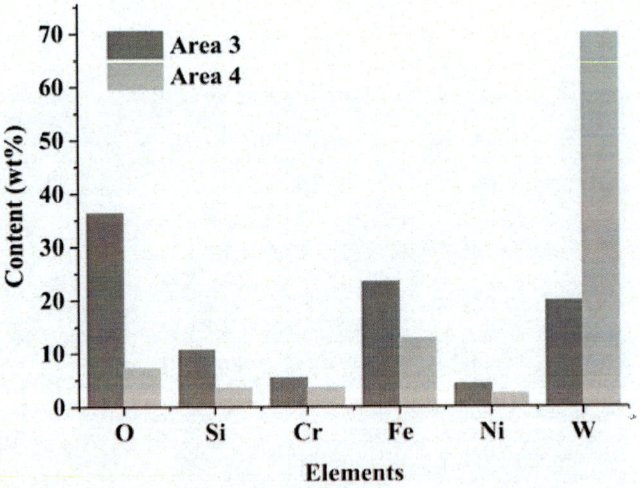

(b)

Fig. 5.18 Element concentrations of EDS analysis on the worn surface of **a** the CS coating and **b** the SLD coating [6]

Stainless steel balls having 8 and 12 mm diameter were used as the grinding media and the ball-to-powder weight ratio was 30: 1. The substrate material was AISI 1045 steel and the substrate specimen had a dimension of 100 × 60 × 10 mm. Before the coating deposition process, the substrate surface was grit-blasted using 24 mesh alumina and ultrasonically cleaned in alcohol.

The same SLD system used in the above studies was used in this research. The gas pressure was 3 MPa, the gas temperature was 500 °C, deposition temperature was 1270 °C, powder feeding rate was 30 g/min, scanning speed was 30 mm/s, standoff distance was 30 mm. For the LC process, a diode laser was used, nitrogen was used as the powder carrier gas and argon was used as the shielding gas. The laser power density was 8.66 × 10^7 W/m^2, the scanning speed was set to 10 mm/s, the thickness of the pre-placed layer of composite powder was 0.8 mm.

5.3.2 *Microstructure of WC/Stellite 6 Coatings*

The microstructural analyses of the obtained coatings were performed with optical microscopy and SEM/EDS/XRD using the same systems described previously. The specimen surfaces were etched with an aqueous solution consisting of 45 mL HCl, 15 mL HNO_3 and 20 mL methanol for facilitating the microstructure observation. In the optical microscopic images of Fig. 5.19, it can be seen that the LC coating contains obvious cracks and pores and the distribution of WC particles is nonuniform

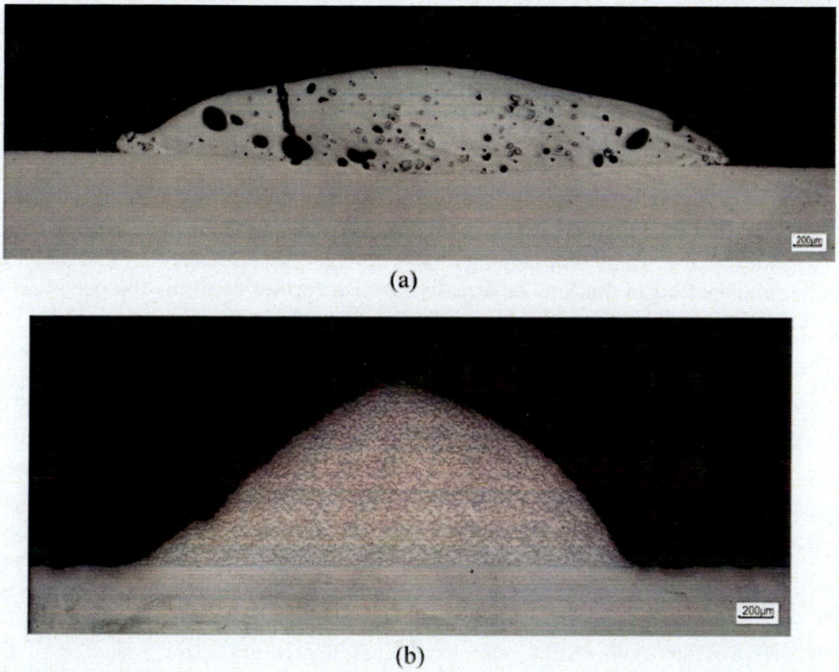

Fig. 5.19 Optical microscopic image in cross-section of **a** LC WC/Stellite 6 coating and **b** SLD WC/Stellite 6 coating [10]

within the Stellite 6 matrix. On the contrary, the SLD coating looks dense and no cracks are observed in the coating, also the WC distribution is more uniform as well.

The formation of pores in the LC process is due to decarburizing reactions in hightemperature molten pool [11]:

$$2WC \rightarrow W_2C + C$$

$$2C + O_2 \rightarrow 2CO, \quad 2CO + O_2 \rightarrow 2CO_2$$

The carbon can dissolve as the WC transforms to W_2C under high laser power, which generates CO_2 and CO gas bubbles in the molten pool. These gas bubbles are subsequently trapped by the rapid solidification of the molten pool leading to the formation of pores in the clad layer. The pores cause stress concentration, allowing easy crack initiation and propagation (Fig. 5.19a). Since the SLD process did not melt the coating and substrate materials, thus no molten pool was formed. As a result, decarburizing reactions of WC can be avoided, allowing to form a continuous and dense coating with minimum pores and cracks (Fig. 5.19b).

The more detailed microstructures of the WC/Stellite 6 coatings were investigated using SEM and the images are shown in Figs. 5.20 and 5.21. The LC coating has a microstructure composed of fine dendrite structures and interdendritic eutectic phases with WC particles embedded in the matrix (Fig. 5.20a). Some of the primary WC particles dissolved into the Stellite 6 matrixforming secondary carbides during the LC process, as indicated by the arrow in Fig. 5.20b. An interface layer between WC particles and the matrix can also be observed, as highlighted by the rectangle in Fig. 5.20b. The formation of this interface should be attributed to the partial melting of the particle surface and inter-diffusion of the particle with the matrix, which formed metallurgical bonding between the WC particle and the matrix. Moreover, from the inset in Fig. 5.20b for mapping of W element, one can clearly see that the W element is dispersed in the Stellite 6 matrix, which further confirms the decomposition of WC particles due to high laser energy in the LC process.

Differently, the SLD coating does not show a dendrite structure. Also, from the EDS mapping analysis of W element (Fig. 5.21a), W element is hardly found in the matrix, indicating no WC decomposition occurring during the SLD process. In the image at higher magnification (Fig. 5.21b), WC particles (bright) and deformed (originally spherical) Stellite 6 particles (dark) are found to be well bonded. These microstructure features reveal that although laser irradiation was introduced, the SLD process still preserved the solidus-state deposition characteristics of CS which can keep microstructural integrity of feedstock materials. The coating formation via SLD depends mainly on the plastic deformation of deposited materials, which is totally different from the rapid melt and solidification in LC.

Similar to the SLD coatings discussed previously, the interface of the SLD WC/Stellite 6 coating is wavy, as shown in Fig. 5.22a, which is attributed to adiabatic shear instability. Taking a WC particle in the SLD coating to examine the interface with the Stellite 6 matrix finds that an interface layer was formed from the

(a)

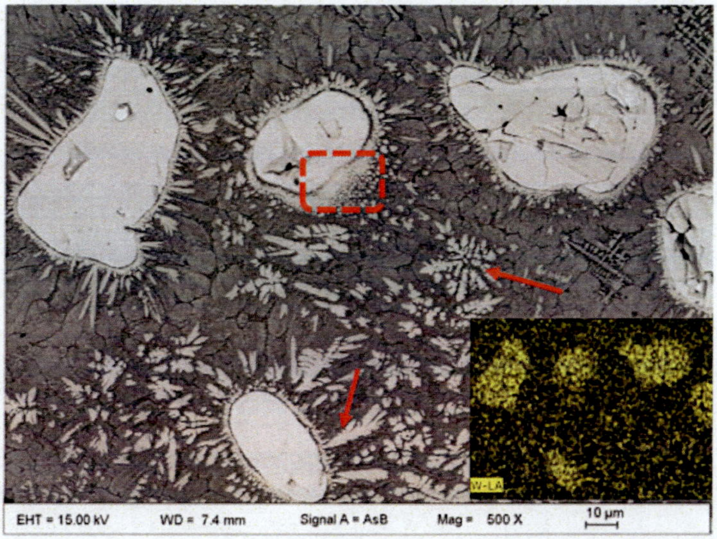

(b)

Fig. 5.20 SEM microstructure of LC WC/Stellite 6 coating: **a** at low magnification and **b** at high magnification [10]

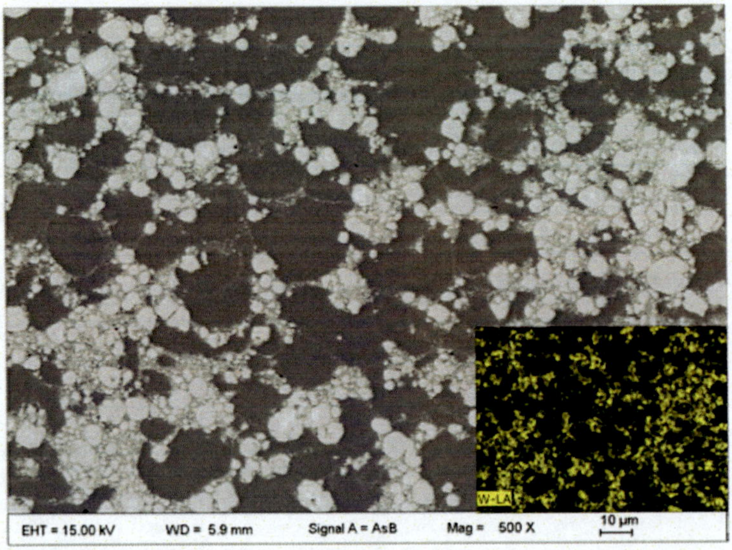

(a)

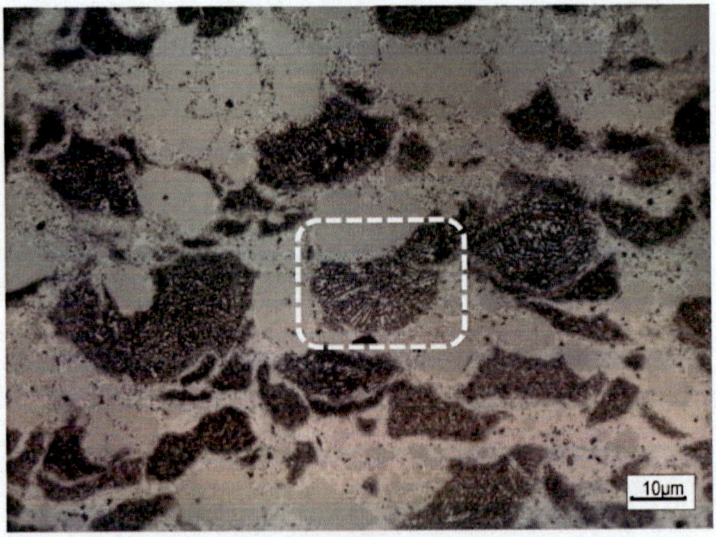

(b)

Fig. 5.21 SEM microstructure of SLD WC/Stellite 6 coating: **a** at low magnification and **b** at high magnification [10]

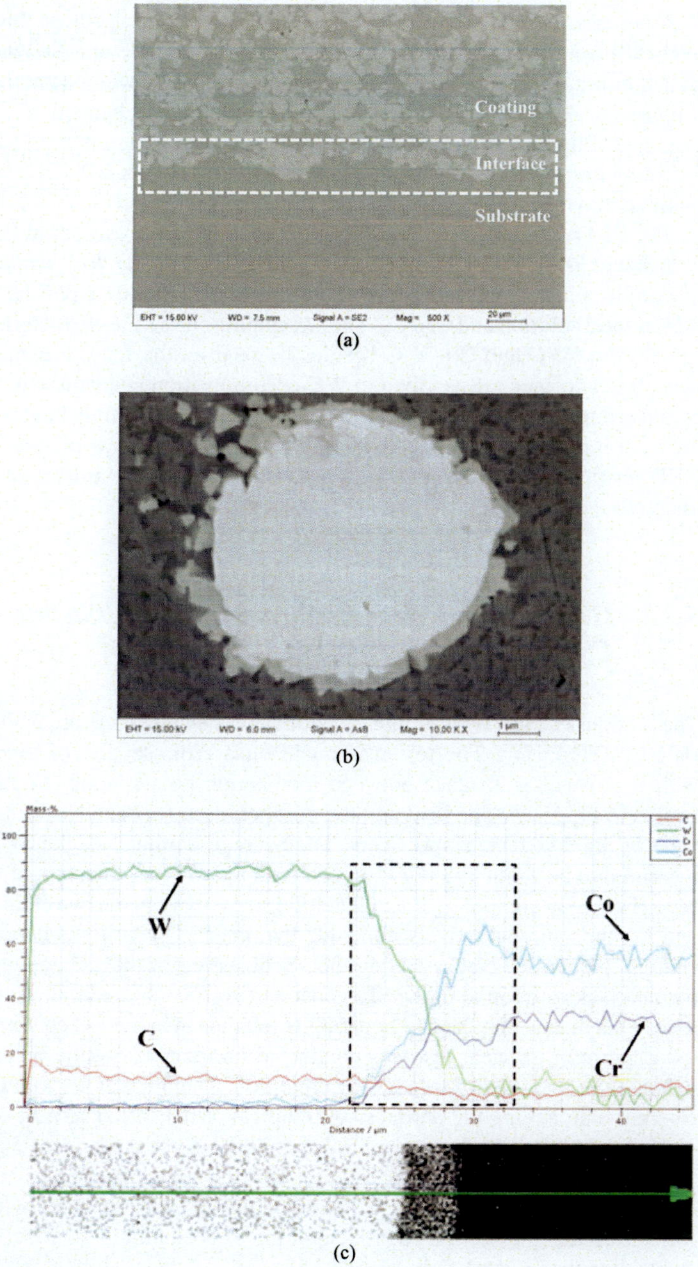

Fig. 5.22 SEM/EDS analyses of SLD WC/Stellite 6 coating: **a** cross-section microstructure, **b** WC/Stellite 6 interface and **c** line scanning across WC/Stellite 6 interface [10]

SLD process, as seen in Fig. 5.22b. The EDS line scanning across this layer reveals inter-diffusion of W, C, Co and Cr between WC particle and Stellite-6 matrix, as shown in Fig. 5.22c, indicating metallurgical bonding developed at the WC/Stellite 6 interface. As discussed above, the main bonding mechanism of CS coatings is mechanical interlocking. Formation of metallurgic bonding in the SLD WC/Stellite 6 coating implies that the introduction of laser irradiation into CS favors interfacial bonding between the WC and Stellite 6 phase.

The XRD patterns of the WC/Stellite 6 coatings are presented in Fig. 5.23, along with that of WC powder and Stellite 6 powder. The pristine WC powder is composed of a single phase of WC without W_2C and the pristine Stellite-6 powder consists of the fcc Co solid solution and $Cr_{23}C_3$. The LC coating shows the characteristic diffraction peaks from W_2C and Co_3W_3C besides those from the feedstock materials. These new phases formed are ascribed to WC decomposition and generation of secondary carbides during the LC process. However, the SLD coating only has the characteristic diffraction peaks from the feedstock materials without new phases formed. These XRD results indicate that the SLD process preserved the phases of the feedstock materials.

5.3.3 Hardness and Wear Performance of WC/Stellite 6 Coatings

The hardness of the WC/Stellite 6 coatings was evaluated using a Vickers Indenter (HR-150DT, DHT). The test was made on a cross-section of the coating along with depth with a distance between indentation points being 0.1 mm, as demonstrated in Fig. 5.24. The load and the duration used for each hardness measurement were 50 kg and 10 s, respectively. The average hardness of the SLD coating was measured to be about 893 HV_{50}, which is slightly higher than that of the LC coating (851 HV_{50}). By comparing the indentation sizes between the two coatings (Fig. 5.24a and b), one can qualitatively conclude that the SLD coating is harder than the LC coating. Moreover, cracks can be clearly observed in the LC coating (Fig. 5.24a), but no cracks are found in the SLD coating (Fig. 5.24b) under the same indentation load. This means that the SLD coating is tougher than the LC coating.

A pin-on-disc wear test was conducted on the WC/Stellite 6 coatings at room temperature under a dry-lubricating condition. The pin was a 4 mm diameter Si_3N_4 ceramic ball, having the hardness of 2200 HV (HRC90). The disk was the coating specimen which was rotating at 500 rpm under a normal load of 500 g during the test. The total sliding time was 120 min for each specimen. The friction coefficients obtained from the wear tests are illustrated in Fig. 5.25, recorded with sliding time. It is shown that the friction coefficient of the SLD coating is smaller (0.62) and more stable than that of the LC coating (0.75).

The worn surfaces of the coating specimens were investigated using a three-dimensional microscopic system with a super depth of field which has the ability to

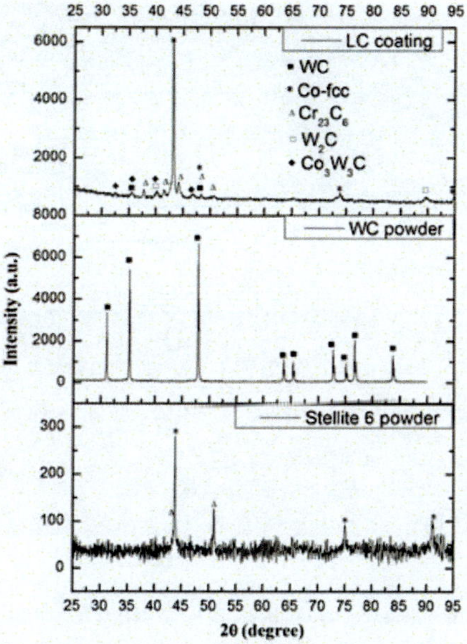

(a)

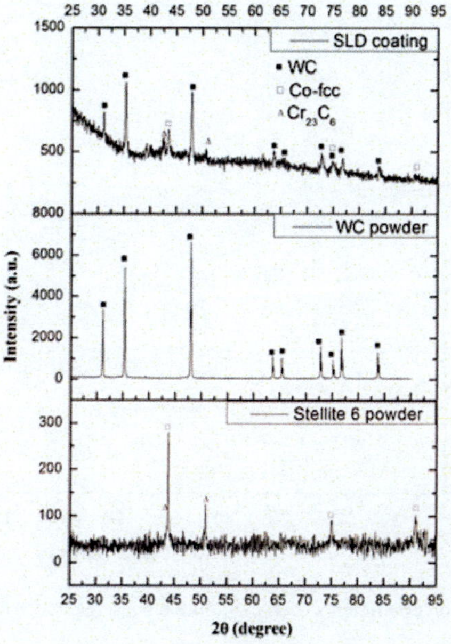

(b)

Fig. 5.23 XRD diffraction pattern of **a** LC WC/Stellite 6 coating and **b** SLD WC/Stellite 6 coating [10]

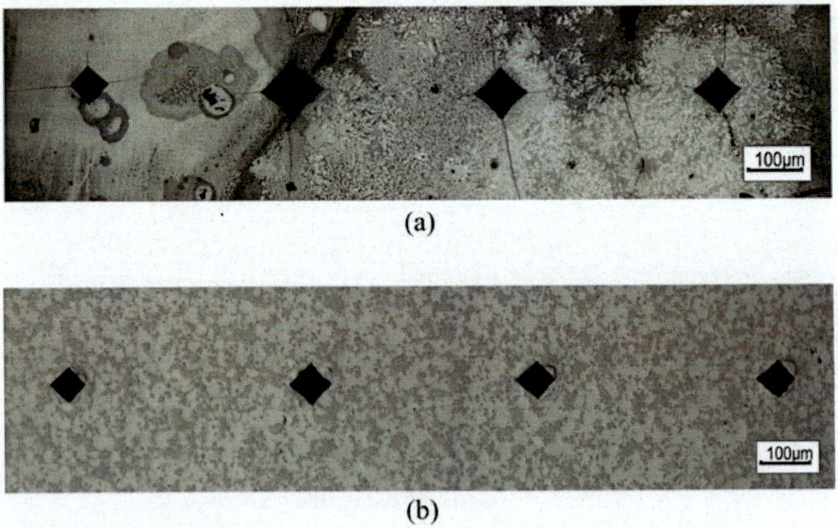

Fig. 5.24 Microhardness test made on **a** LC WC/Stellite 6 coating and **b** SLD WC/Stellite 6 coating [10]

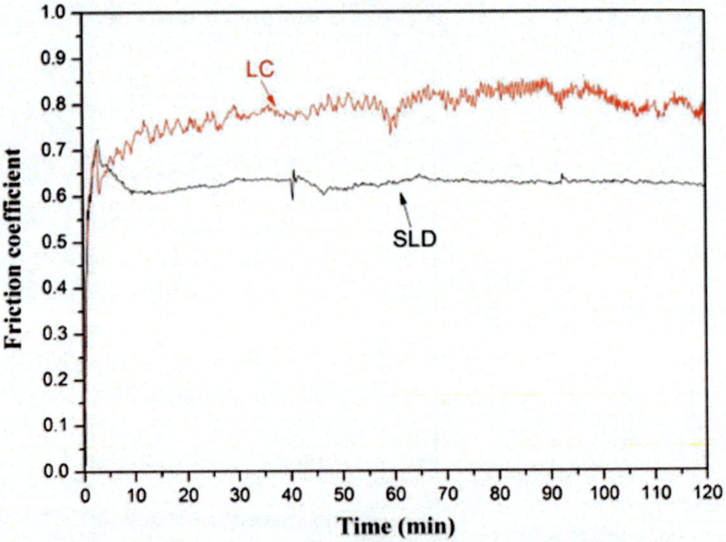

Fig. 5.25 Friction coefficients of WC/Stellite 6 coatings under dry-sliding wear [10]

measure the width and depth of a wear track and thus to evaluate the volume loss or wear loss. The morphologies of the worn surfaces of the coating specimens are shown in Fig. 5.26 and the measured depths and calculated volumes of the wear tracks are reported in Fig. 5.27. It is seen that the wear track width of the SLD coating is about 525 μm, which is smaller than that of the LC coating (625 μm). Also, the maximum wear track depth of the SLD coating is about 4.7 μm while that of the LC coating is around 23 μm. Accordingly, the volume loss of the SLD coating is much less than that of the LC coating. Therefore, it can be concluded that the WC/Stellite-6 composite coating prepared by SLD has superior wear-resistant performance to the counterpart prepared by LC, which is attributed to the fact that SLD can result in fewer defects, higher WC content, lower cracking susceptibility of the coating. SLD is also able to preserve the microstructure and phase composition of the coating feedstock material due to relatively low heat involved in the process. Moreover, because of the additional heat provided by laser irradiation, the bonding mechanism of the SLD coating evolves from dominant mechanical bonding (CS coating) to coexistence of mechanical and metallurgical bonding.

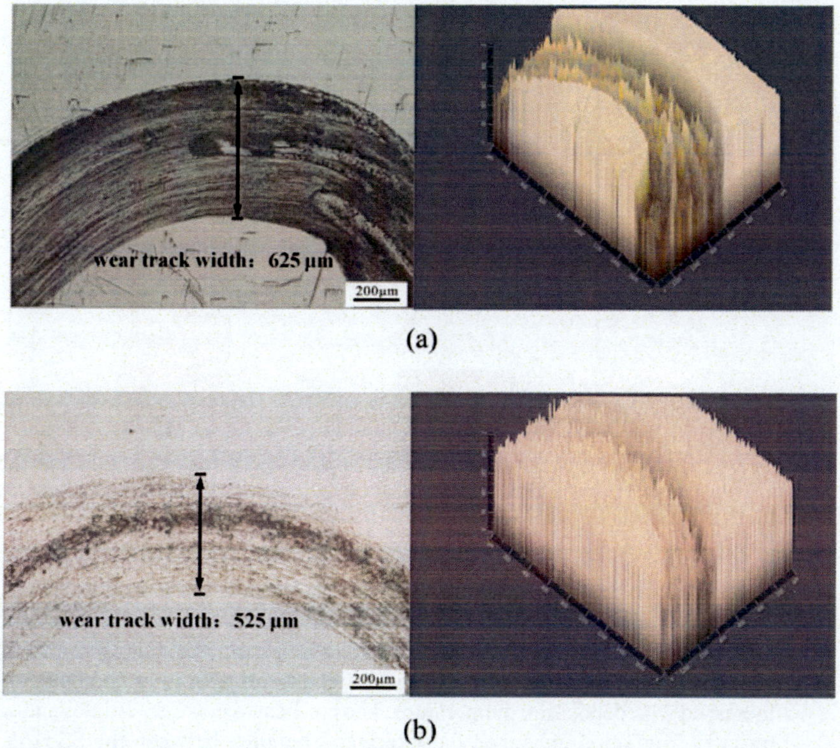

Fig. 5.26 Wear track morphology and 3D profile of **a** LC WC/Stellite 6 coating and **b** SLD WC/Stellite 6 coating [10]

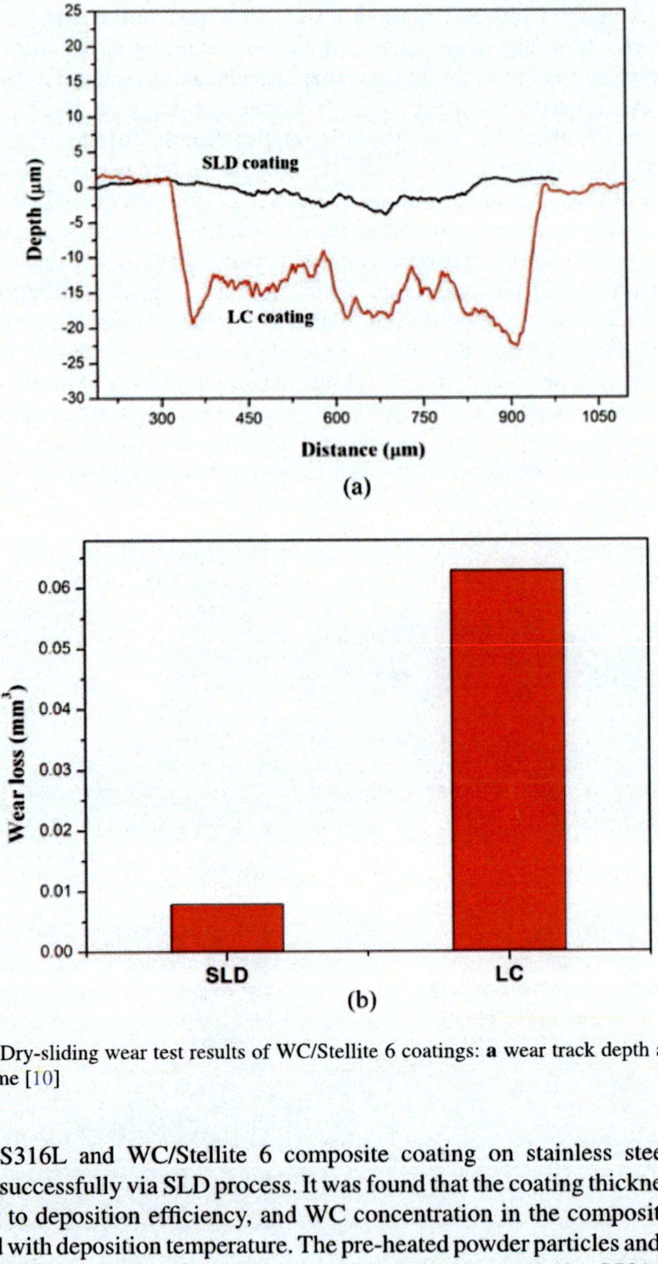

Fig. 5.27 Dry-sliding wear test results of WC/Stellite 6 coatings: **a** wear track depth and **b** wear track volume [10]

WC/SS316L and WC/Stellite 6 composite coating on stainless steel can be obtained successfully via SLD process. It was found that the coating thickness, corresponding to deposition efficiency, and WC concentration in the composite coating increased with deposition temperature. The pre-heated powder particles and substrate by laser irradiation enhanced the interfacial bonding between the SS316L matrix and WC particles and the interfacial bonding between the coating and substrate. The SLD process did not cause the change of chemical composition and phases in the WC/SS316Lcomposite coating, compared with the feedstock powder. The high content of WC and strong interface bonding of the particles dramatically contribute to the tribological properties of the composite coating.

The WC/Stellite 6 coating fabricated via SLD appeared continuous and dense without obvious defects such as pores and cracks. It preserved the microstructures and phase compositions of the feedstock materials due to the relatively low temperature solidus-state deposition process. As a consequence, the SLD coating had higher WC content, lower cracking susceptibility, better tribological performance, compared with the LC coating. Furthermore, laser irradiation enabled creating of metallurgical bonding in the SLD coating.

5.4 Summary of Supersonic Laser Deposition of Composites

As an improved coating deposition approach over traditional cold spray (CS), supersonic laser deposition (SLD) process makes it possible to deposit hard materials based on CS. In addition to successfully depositing Ni60 and Stellite 6 alloy from SLD processes, hard composite coatings have also been achieved via SLD.

Two different sizes of diamond particle mixed with hard Ni60 powder were deposited on medium carbon steel using SLD process. The pre-heated powder particles and substrate by laser irradiation enhanced the interface bonding between the coating and substrate, and between the Ni60 matrix and diamond particles. It was found that the diamond particles were uniformly distributed in the Ni60 matrix and the coating was dense without visible cracks. The diamond concentrations in the two composite coatings were similar. The Raman typical peaks of diamond particles in the two composite coatings were consistent with that of diamond material, showing non-graphitization transition occurred during the coating deposition process. Both coatings exhibited dense microstructure and excellent wear resistance. The Ni60 matrix was strain-hardened by the repeated impacts of diamond particles in the SLD process. The smaller-size diamond was better than the larger-size one in the composite coatings for wear resistance.

A comparison between the WC/SS316L composite coatings produced with CS and SLD was made with respect to deposition efficiency, WC distribution and concentration, interfacial bonding, phases in the microstructures, and tribological properties as well. The experimental results showed that with the assistance of laser irradiation the deposition efficiency, WC concentration and interface bonding of WC/SS316L composite coating were greatly improved, compared with the coating prepared with CS. The SLD coating had the same phases as the CS coating owing to the low heat involved in the SLD process. The high concentration and strong interface bonding of WC particles in the SS316L matrix dramatically improved the tribological properties of the SLD coating.

The application of SLD to producing WC/Stellite 6 composite coating demonstrated the solid-state manufacturing of continuous and dense coating. The comparison between the WC/Stellite 6 composite coatings prepared by SLD and laser cladding (LC) was conducted with respect to microstructure, phase composition, microhardness, cracking susceptibility and tribological performance. The experimental results revealed that the SLD coating had fewer defects, higher WC content, lower cracking susceptibility and better wear resistance than the counterpart prepared via LC, which was attributed to the preservation of microstructure and phase compositions of the feedstock materials due to relatively low heat involved in the SLD process. On the other hand, because of the additional heat provided by laser irradiation, the bonding mechanism in the SLD coating evolved from dominant mechanical bonding in the CS coating to coexistence of mechanical and metallurgical bonding.

References

1. M. Iravani, A. Khajepour, S. Corbin, S. Esmaeili, Pre-placed cladding of metal matrix diamond composite on mild steel. Surf. Coat. Technol. **206**, 2089–2097 (2012)
2. J. Miksovsk, A. Voss, R. Kozarova, P. Pisarik, G. Ceccone, W. Kulisch, M. Jelinek, M.D. Apostolova, J.P. Reithmaier, C. Popov, Cell adhesion and growth on ultrananocrystalline diamond and diamond-like carbon films after different surface modifications. Appl. Surf. Sci. **297**, 95–102 (2014)
3. W. Li, Y. Wang, S.H. Fan, J.F. Xu, Wear of diamond grinding wheels and material removal rate of silicon nitrides under different machining conditions. Mater. Lett. **61**, 54–58 (2007)
4. L.J. Yang, B. Li, J.H. Yao, Z.H. Li, Effects of diamond size on the deposition characteristic and tribological behavior of diamond/Ni60 composite coating prepared by supersonic laser deposition. Diamond Related Mater. **58**, 139–148 (2015)
5. N.M. Melendez, A.G. McDonald, Development of WC-based metal matrix composite coatings using low-pressure cold gas dynamic spraying. Surf. Coat. Technol. **214**, 101–109 (2013)
6. B. Li, J.H. Yao, Q.L. Zhang, Z.H. Li, L.J. Yang, Microstructure and tribological performance of tungsten carbide reinforced stainless steel composite coatings by supersonic laser deposition. Surf. Coat. Technol. **275**, 58–68 (2015)
7. H. Assadi, F. Gartner, T. Stoltenhoff, H. Kreye, Bonding mechanism in cold gas spraying. Acta Mater. **51**, 4379–4394 (2003)
8. M. Grujicic, J.R. Saylor, D.E. Beasley, W.S. DeRosset, D. Helfritch, Computational analysis of the interfacial bonding between feed-powder particles and the substrate in the cold-gas dynamic-spray process. Appl. Surf. Sci. **219**, 211–227 (2003)
9. C.J. Li, W.Y. Li, Y.Y. Wang, Formation of metastable phases in cold-sprayed soft metallic deposit. Surf. Coat. Technol. **198**, 469–473 (2005)
10. B. Li, Y. Lin, J.H. Yao, Z.H. Li, Q.L. Zhang, Solid-state fabrication of WC_p-reinforced Stellite 6 composite coatings with supersonic laser deposition. Surf. Coat. Technol. **321**, 386–396 (2017)
11. C.P. Paul, H. Alemohammad, E. Toyserkani, A. Khajepour, S. Corbin, Cladding of WC-12Co on low carbon steel using a pulsed Nd:YAG laser. Mater. Sci. Eng. A **464**, 170–177 (2007)